Mapping: Ways of Representing the World

Daniel Dorling
Department of Geography, University of Bristol

and

David Fairbairn
Department of Geomatics, University of Newcastle upon Tyne

An imprint of **Pearson Education**

Harlow, England · London · New York · Reading, Massachusetts · San Fran
Tokyo · Singapore · Hong Kong · Seoul · Taipei · Cape Town · Madrid · Me

D1416599

Pearson Education Limited
Edinburgh Gate
Harlow
Essex CM20 2JE
England

and Associated Companies throughout the world
http://www.pearsoneduc.com

First published 1997

ISBN 0 582 28972 6

British Library Cataloguing-in-Publication Data
A catalogue record for this book is
available from the British Library.

Library of Congress Cataloging-in-Publication Data
A catalog entry for this title is
available from the Library of Congress.

Set by 34 in 11/12 pt Adobe Garamond

Printed in Great Britain by Henry Ling Ltd., at the Dorset Press,
Dorchester, Dorset.

10   9   8   7   6   5   4   3
05   04   03   02   01

Contents

Foreword

This volume is part of a series of new human geography teaching texts. The series, *Insights in Human Geography*, is designed as an introduction to key themes in contemporary human geography. Together, the volumes form the basis for a comprehensive approach to studying human geography. Each volume, however, is free-standing and can be studied on its own as an introduction to a specific sub-field within the discipline.

The series is built around an exploration of both the substantive geographies of the real world and the conceptual and theoretical frameworks that are required to contextualize them. Each volume not only provides a thorough grounding in one particular sub-field of human geography but also looks to the broader picture, providing students with a geographical perspective on contemporary issues, and showing how recent changes in the real world have led to changes in the ways that geographers approach and understand the world.

This volume, *Mapping: Ways of Representing the World*, takes as its subject-matter one of the fundamentals of geographic scholarship and research. Map-making and the interpretation of maps, though they have long been acknowledged as basic elements in geographic approaches to understanding the world, have often been approached in rather narrow ways. As a result, many geographers (and most other observers) developed a taken-for-granted attitude towards mapping and map-making, or else saw the field as one of arcane technicalities and refined aesthetics.

This volume shows that there is much more to mapping, map-making, and the interpretation of maps than technical and aesthetic considerations. By bringing together ideas from cultural geography, cartography and geographical information systems, the authors encourage students to think about *how* and *for whom* the world is mapped. In addition to an explanation of the fundamentals of mapping and map-making, the book shows how maps tell as much about the people and powers that create them as about the places they depict.

Acknowledgements

We would like to thank Professor Jeremy Black, the series editors and Sally Wilkinson of Addison-Wesley Longman for their constructive comments on the first draft of this book.

D.D. would like to acknowledge the assistance and encouragement of colleagues in the Geography Departments of the Universities of Newcastle upon Tyne and Bristol; members of the ICA Commissions on Multimedia, Education and Training, and Map Use; the ICA Working Group on temporal issues in GIS; and Bronwen Dorling for helpful comments.

D.F. would like to acknowledge academic assistance from colleagues in the (previously named) Department of Surveying, University of Newcastle upon Tyne; the Department of Land Information, Royal Melbourne Institute of Technology University; and the Map Library of the State Library of Victoria, Melbourne; and support from Sue and the children.

We are grateful to the following for permission to reproduce copyright material: Figure 1.3 from *The History of Cartography*, Vol. 1, University of Chicago Press (Harley & Woodward, eds.); Figure 3.1a from part of the monochrome *London Transport Underground Map* (area from Shepherds Bush to Euston Square (west–east), and Bond Street to Camden Town (south–north), LRT Reg. User No. 97/2482, London Transport Museum; Figure 3.2 from a portion of AAA's triptik, American Automobile Association © AAA reproduced with permission; Figure 3.4 from *Great Ormes Head to Liverpool* (1978) and Figure 3.5 reproduced form *Admiralty chart 1978*, 1795–1995 leaflet, © Crown copyright, reproduced by permission of the Controller of HMSO and the UK Hydrological Office; Figure 5.2 from Ordnance Survey OS 1:50,000 map, extract Crown Copyright; Figure 6.3 from *Cartographic Relief Presentation* edited by H. J. Steward, Walter de Gruyter (Eduard Imhof, 1982); Figure 9.3 (p. 6 fig. 1.3) and Figure 9.4 (p. 7 fig. 1.4) reprinted from *Visualization in Modern Cartography*, with kind permission from Elsevier Science Ltd, The Boulevard, Langford Lane, Kidlington OX5 1GB, UK (MacEachren & Taylor, eds., 1994).

Whilst every effort has been made to trace the owners of copyright material, in a few cases this has proved impossible and we take this opportunity to offer our apologies to any copyright holders whose rights we may have unwittingly infringed.

Introduction

Geography and mapping

The role of geography is to analyse and explain the phenomena of the landscapes that surround us, as well as to monitor the nature of human society and the economic, social, political and cultural lives that we live. Although these two strands might be regarded as requiring the investigation of mutually exclusive factors and issues, they can both be approached from a geographical perspective because of their shared *spatial* nature. It is partly this characteristic which distinguishes geography from other physical and social sciences, and although the primary means of geographical discourse is textual, the tools which geographers use to undertake their tasks do include those which explicitly reflect the spatial dimension of their field of knowledge.

Maps have fulfilled the role of spatial data handling tools for millennia, considerably longer than the period in which a study named 'geography' has been recognized or taught. Despite its relative youth, however, geography has been characterized by prevalent fashions and the changing paradigms within which it is studied, by its changing methods of investigation and by changing perceptions of its function and purpose: shifts in emphasis to which, until very recently, maps appeared relatively immutable. This book is intended to reflect on the nature of mapping and map-making in the era before formal geographical study; to discuss the methods of map production and the directions of map use, particularly in geography but also in many other spheres of human activity which use maps, in more modern times; to consider the (now) rapidly changing nature of cartography; and to speculate on the future direction of mapping and its relationship with geography. Throughout the book a question which is continually raised is how the creation and use of maps affects the various ways in which we think of and view the world.

Being useful and already well-established tools, maps, as representations of the space we inhabit, were naturally attractive to human geography as an infant discipline. Initially conceived as the study of regional variation in human activity throughout the world, human geography called upon cartographers to describe, in graphical terms, diversity on the surface of the earth. Regional geography had its roots in ancient study and voyages of exploration and discovery of 'different' places. Explanations of the diversity encountered needed maps showing environmental features in particular, and the pre-eminence of environmental determinism in the late nineteenth century was strengthened by the use

1

of such maps showing human activity being governed by associated physical features in their locality. Even when the role of human agents in determining the relative locations and distributions of social and economic activity became recognized as important, by early twentieth-century geographers such as Sauer and de la Blache, the map was still regarded as the tool *par excellence* for conveying geographical descriptions and explanations – the extent of the discipline of geography at the time. Building on the regional approach, British geographers, such as Geddes and Stamp, used maps, often created as a result of intensive fieldwork, in planning and civic applications.

Beyond mere description, post-war geographers up to the 1960s attempted to discern order and establish rules governing spatial behaviour. Such universal laws were to be obtained by scientific investigation and to be rendered using mathematical and geometric language in a positivist manner. The neutrality of the observations undertaken by such 'spatial scientists' relied on the notion of closed geographical systems within which human beings operated, and the constrained nature of the map sheet and its stylized representation of reality proved the ideal source for initiating and confirming hypotheses.

The reaction against looking purely at the closed systems represented by maps was led by researchers labelled as Marxist geographers in the 1970s, who felt that spatial processes and resultant patterns required economic, social and political explanations; and by humanist geographers of the 1980s who recognized the importance of the subjective nature of geographical investigation, both in terms of the observer and the observed. A humanist view is that the nature of data collection and representation, including mapping, is likely to affect the results of geographical enquiry. Further contemporary geographical thought is based on ideas of post-modernism, wherein the notion of predictable spatial order and of a 'grand design' that can be mapped, understood, explained and predicted, is not acceptable. Indeed for some geographers, any form of scientific precision is regarded with suspicion, replaced by a recognition of the value of opinion, interpretation and 'hidden agendas'.

These diverse views of the nature and meaning of geography reflect its increasingly pluralist tradition. Reflecting contemporary society (as it always has), geography is becoming a highly disparate field and the prevalence of one particular paradigm is being replaced by a multiplicity of views regarding the essence of what geography is. In addition to the succession of persuasions indicated above, feminist, social theoretic and structurationist viewpoints in geography each make different assumptions about what the world is like: mapping is related to such debates, disagreements and different conceptualizations.

This lack of coherence in geography masks a fundamental observation that, although maps are viewed differently by different geographers, mapping is a vital geographic technique of study and maps are a primary tool. Some geographers (notably Hartshorne) have, in the past, considered that maps define the very nature of geography. All schools of geographic thought call attention to mapping as a geographic method. Indeed, it is so important to geography that mapping as a technique and the map as a tool are often the most critically analysed features when paradigm shifts occur in geography. Unfortunately, as indicated by Krygier (1996), this interest is not necessarily reciprocated by those whose primary role is to prepare maps: 'cartographers have had little to say about how [the] fundamental differences in geographic philosophy and theory relate to maps ... nor have cartographers addressed the issue of why certain approaches to geography tend to use maps and other visual representations more than others' (Krygier, 1996, p. 25).

This inevitably brief overview of the development of human geography has shown how it may have varying reliance on maps as products to be created and used. What characterizes the map, however, is its unique role in the geosciences and outside the academy in general, and the intention of this book is to examine the significance of mapping and map-making and their contribution to the study of human activity (including human geography) and, by drawing on numerous examples, to peoples' everyday lives around the world.

Maps as the subject of this book

In order to examine the nature of maps, the book reflects on the ways in which societies around the world have interpreted and represented the world on which we live. Mapping is only one of these techniques, but is a fundamental form of representing space and location, and it occupies a central place in the experience and development of the majority of people who have lived in large groups. Just as societies and civilizations vary enormously in their essential characteristics, so their map representations vary – in purpose, scale, content and conceptions of accuracy. The physical creation of maps which embody the 'world-view' of such societies is the process of *map-making*. This can be distinguished from the mental interpretation of the world which is termed *mapping*. Clearly the latter must precede the former, although the act of mapping may not result in the production of a map artefact. The *map* product itself can be defined as 'a symbolized image of geographic reality, representing selected features or characteristics, resulting from the creative efforts of cartographers and designed for use when spatial relationships are of special relevance' (International Cartographic Association, 1995, p. 1). This deliberately broad definition encompasses not only current familiar and official maps, which possess such rigorous cartographic components as a scale, legend and reference system (a grid or graticule), but also a large range of *map-like objects*, a term given to many (other) spatial representations, both those of pre-history and contemporary sketch maps and diagrams. Despite lacking the features, such as a grid, which are necessary to undertake many map-use tasks, the latter still have as their function the representation of the space that we inhabit. Unconventional maps often also convey different types of information from that embedded in what has become the traditional map – information about the spatial organization of nature and society which is usually concealed. Although diagrammatic and caricature maps are included in the definition of maps used here, this definition stops short of including such illustrations as chess drawings and scatter-plot graphs.

Because the study of mental representations is fraught with difficulty, geographers have tended in the past to focus their study of cartography ('the discipline dealing with the conception, production, dissemination and study of maps in all forms': International Cartographic Association, 1995, p. 1) on the products themselves, describing their variation and their methods of production. Unfortunately, therefore, cartography has tended to be regarded merely as a technology, and its practice has been seen as an acquired skill, capable of being undertaken by an artisan, rather than a scientist, and definitely not an everyday activity. The contemporary study of cartography, however, is widening to encompass all aspects of mapping. This is due to the increasing applications of cartography in a wide range of earth and social sciences, the recognition of the core place of cartography in current areas of development in information technology (such as geographical information systems, visualization, multi-media and virtual reality), and the recognition that the craft skills required to undertake map-making are diminishing in importance. A renewed interest in cartography is evident in some of the schools of geographical enquiry considered above. The vibrancy of cultural geography especially, which deals with the relationship between humankind and the territory in which it leads its active existence, acknowledges the utility of maps in describing and analysing that relationship, along with the fundamental nature of mapping and map-making as human activities.

The relationship between humans and maps is worthy of study: the map is a store of knowledge that can 'add value' to human activity. It can be used to order the territory itself, as well as to order our knowledge of it. Our perception of the world is constantly being moderated by our experiences of mapping, map-making and map using. Maps are continually used to interpret and present information about the real world. Although a scientific view would suggest that such information be portrayed in a neutral, objective, impersonal, unadorned manner, and that maps disengage us from a personal, subjective view of the world, this book will argue that maps are not independent of the view of the observer: maps are context-dependent, often available only to the initiated, unlikely to be value-free and should be viewed with caution (although not necessarily scepticism). The role of the human being in interpreting and rendering world-views in map form is central.

Further reading

D. Livingstone's masterly survey of the history of geographical thought, *The Geographical Tradition: Episodes in the History of a Contested Enterprise* (Blackwell, Oxford, 1993), indicates the importance of pre-nineteenth century mapping and exploration activity for the embryonic development of geography as a discipline. J. Krygier gives a cartographer's view of the changing nature of geography and its impact on mapping and map-making activity in 'Geography and Cartographic Design', Chapter 3 in *Cartographic Design: Theoretical and Practical Perspectives*, edited by C. H. Wood and C. P. Keller (Wiley, Chichester, 1996). The strands of twentieth-century and contemporary geographic thought are brought together in *Approaching Human Geography: An Introduction to Contemporary Debates*, by P. Cloke, C. Philo and D. Sadler (Chapman, London, 1991).

The discipline of cartography has been well served by written works throughout its long history, and there are important and relevant overviews which should appear on any reading list associated with mapping and map-making. The links with techniques of geographical data collection and measurement are covered in works as diverse as *Surveying and Mapping for Field Scientists* by W. Ritchie, D. Tait, M. Wood and R. Wright (Longman, Harlow, 1988) which describes elementary land and aerial survey techniques, and *The Census Users' Handbook*, edited by S. Openshaw (GeoInformation, Cambridge, 1995) which considers socio-economic data collection and GIS: see, in particular, Chapter 5 by M. Charlton, L. Rao and S. Carver. The importance of presentation is stressed in the major cartographic textbooks: J. Keates' *Cartographic Design and Production* (2nd edition, Longman, Harlow, 1989); *Elements of Cartography*, now in its 6th edition, authored by A. Robinson, J. Morrison, P. Muehrcke, A. Kimerling and S. Guptill (Wiley, Chichester, 1995); and a valuable new work by M. Kraak and F. Ormeling, *Cartography: Visualisation of Spatial Data* (Longman, Harlow, 1996). Examples of manuals of 'how to' present maps include *Some Truth with Maps* by A. MacEachren (Association of American Geographers, Washington, 1994) and *Cartography: Thematic Map Design* by B. Dent (Wm. Brown, Dubuque, Iowa, 1993).

A number of books by M. Monmonier give thoughtful and incisive accounts of the role and potential of cartography in geographical analysis and social enquiry: *How to Lie with Maps* (University of Chicago Press, Chicago, 1991); *Mapping it out: Expository Cartography for the Humanities and Social Sciences* (University of Chicago Press, Chicago, 1993); and *Drawing the Line* (Henry Holt and Company, New York, 1995). Further accounts of the nature of cartography and its place in the social sciences can be found in *How Maps Work*, by A. MacEachren (Guilford, New York, 1995), a detailed exploration of the psychological and physiological concerns of the cartographer and map user; *Understanding Maps*, by J. Keates (2nd edition, Longman, Harlow, 1996) which addresses similar concerns regarding map creation, map production and map use; and *The Power of Maps*, by D. Wood (Guilford, New York, 1992) which examines the 'post-modern' agenda of contemporary cartographic thought. The current increase in interest in cartographic practice and its possibilities in the handling of spatial data in the digital environment have led to new analyses of the impact of accepted cartographic customs (exemplified by Chapter 2, 'The Traditional Map as a Visualization Technique', by M. Wood in *Visualisation in Geographical Information Systems*, edited by H. Hearnshaw and D. Unwin, Wiley, Chichester, 1994), the place of cartography in GIS (see Chapter 3, 'Visualization in GIS, Cartography and ViSC', by M. Visvalingam, also in Hearnshaw and Unwin, 1994), the changing nature of cartography's traditional customers (see in particular, Chapters 11–16 in *Visualization in Modern Cartography*, edited by A. MacEachren and D. Taylor, Pergamon, Oxford, 1994) and the possibilities of 'new' cartography with a different outlook on the presentation of spatial information (see *A New Social Atlas of Britain*, by D. Dorling, Wiley, Chichester, 1995).

Chapter 1

The history of cartography

An introduction to early maps

The Introduction to this book suggested that different peoples and societies can have very different perceptions of the world, and that there exist, therefore, a variety of ways of interpreting and mapping it. If maps result from such interpretations, they too will vary, and reflect their makers' 'world-views'. Imbued with meaning, inference and prejudice, both conscious and subconscious, it is clear that maps are not simple representations of reality. The most effective way of understanding this is to study ancient maps, including map-like objects, looking, as far as we can now determine, at what their creators chose to show, how it was shown and what embellishments were made on the map face. The world-views that are reflected in contemporaneous map production vary from one cartographer to another and from one culture to another. Throughout history, personal and societal influences have been as strong as the shape of the landscape in determining the appearance and content of maps.

Virtually all early map products attempted to represent impressions of the landscape. In many cases the 'landscape' of the earth was placed in its assumed position in the cosmos: many maps were (and many still are) representations of theories and views of the universe. Alternatively, early cartographers often attempted to portray a view of the entire earth as they saw it, concentrating on the distribution and relationships of known or imagined features within the bounds of the planet. Such small-scale map products were balanced by occasional large-scale representations which covered considerably smaller tracts of landscape and which were created for more specific purposes, such as navigation or social regulation.

It must be appreciated that the study of such maps can encompass an enormously long time period (the earliest surviving map artefacts date from 3500 BC, although prehistoric rock art, some of which may have 'proto-cartographic' features, dates back to the Palaeolithic period of 30 000 BC) and embraces a vast range of cultures and societies, each with its own ideas about how its world-view could be represented in map form. Differing cultures had differing inclinations to produce such varying maps. Prehistoric peoples lived closer to nature, were dependent upon it and often relied on mobility to follow a hunter–gatherer style of life. Their innate capacity to create a coherent reference framework from the

6

natural surroundings ('spatialization'), their well-developed sense of vision and the sketching ability demonstrated in the cave paintings which still exist, led in the majority of cases to the development of map-making as a means of communicating and recording information. Later, more settled societies used local maps for inventory and management, but still produced maps based on world-views for displaying the place and nature of the earth.

The human mind and the shape of the earth: reconciling interpretation and reality

In many societies there were irreconcilable differences in the world-view shown by maps produced by and for the philosopher or thinker, and those of the traveller or scientist. Variation between cultures has also been apparent, even at similar time periods in history. For example, the significantly different traditions emanating from Greece and from Rome during a common period in the two centuries around the time of Christ show the wider intellectual view of the Greeks contrasted with the more practical measurement-based cartography of the Romans. The former was much influenced by philosophical reflection, but also by astronomical observation, and resulted in often speculative world maps (see p. 9). The latter took a considerably more structured technological approach to the measurement of distance and angles on the ground, to allow for the creation of large-scale land ownership maps and route diagrams. Roman maps depicting the entire world were rare (Figure 1.2) and displayed a limited perception of areas beyond their own, measurable, realm.

Such variation reveals that the development of map-making has often been fitful: there are enormous historical gaps in any perceived 'progression' and many societies felt perfectly able to function without any formal map-making output at all. It is thus difficult to put forward a coherent, generalized view of the history of cartography (see Box 1.1 for an outline of the paradigms within which the subject can be studied). However, it can be suggested that, prior to the period characterized as 'The Age of Discovery' (starting about AD 1490), world and cosmos maps tended to be generalist, philosophical interpretations of religious and superstitious belief whilst local maps were much more practical, used for land appropriation or management and other specific purposes. From the turn of the sixteenth century, a more enlightened approach to cartography was apparent as observation and measurement became the foundations of map-making, although at certain previous times, within Chinese cultures in particular, such practices were prominent. It should be recognized that this more 'scientific' view of the mapping process was, and still is, tempered by the prevalent ideology of the map-maker, who is able to manipulate the appearance and content of the map to a surprisingly large degree. The reconciliation of an interpretation of reality (and its subsequent representation in map form) with reality itself may be satisfactorily carried out only for some map users and some map-use tasks, not for all: maps are not mirrors of reality and the mind of their creators is imprinted on every map.

Box 1.1   Contemporary mapping box – Methods of studying the history of cartography

There are certain approaches that can be made in the study of the history of cartography, ranging from that of the 'interested amateur', who might be inclined to visit book fairs and antique shops in the hope of discovering a valuable old document, to the analytical study of the serious historian, who may try to rationalize the appearance and content of a map by detailing the various factors that affected the mapping and map-making processes.

There is a distinctively cartographic viewpoint which combines a study of the map documents, the technologies which allowed their creation, and an appreciation of the context within which mapping activity developed. Within this approach there are a number of different 'paradigms' or frameworks of study:

- The 'Darwinian' view: map-making improves as civilization progresses, and knowledge of spatial data moves 'from myth to map'. Here, in particular, accuracy is examined – accuracy of geodetic and planimetric systems, and also of content and representation.
- The 'old is beautiful' view: most cartographic research has been undertaken on older maps. Maps from the Renaissance have been studied much more fully than those of the early twentieth century, for example.
- The 'nationalistic' view: concentrating on the map production in one area or one nation state is also seen as a valid way of following the development of map-making.

Although accuracy is a prime measurable indicator of 'progress' in map-making, there are other map elements which can be considered:

- the narrow meaning of maps and their graphical rendering – their use as documents to communicate specific messages to the reader;
- the use of maps in a wider societal context, in particular as tools of oppression, governance, policy-making and regulation;
- the intellectual endeavour required to create and reproduce maps, looking at mapping from viewpoints as diverse as psychological investigations into ancient views of the earth and the history of the technology required to print and disseminate maps;
- the artistic representation on the map face, reflecting on the map as a decorative *objet d'art* or an adjunct to artistic output.

The relationship of mapping to other human activities

The history of cartography has usually been written from a chronological perspective (see Further reading) often within the 'Darwinian' paradigm described in Box 1.1. This chapter, however, selectively considers periods in the history of cartography which exemplify the connections between map-making and other societal activities. The propensity to map is a function of such activities, which include the accumulation of spatial data (from philosophical insight, religious belief and scientific observation), the recognition of the utility of spatial representation (in navigation, education, land management, and for political ends) and the ability to create map products (using contemporary technology).

It is important to note also that this chapter, along with most other works on the history of cartography, does tend to take a Eurocentric view of the history of cartography. This reflects the rich inheritance of surviving map artefacts from western cultures which has tended to dominate the interest of historians of cartography, and occurs despite the numerous encounters by western explorers and map-makers with indigenous peoples which have given insight into the mapping urges of others. There has been, at certain stages of history, an evident lack of map-making activity in some societies outside the 'western' realm (as well as within it), but this may not imply a lack of mapping ability. Indeed, mapping ability (as distinct from map-making) seems innate in every culture.

Philosophy and its influence on mapping

Speculation on the place of the earth within a cosmic framework was one of the most important roles for early philosophers. These views developed in those ancient civilizations with sufficient specialization of labour to allow for the establishment of groups of such scholars. The world-views propounded by them varied enormously, viewing the earth as, for example, a square (early Chinese philosophy in the Han dynasty up to the second century BC), as one of a number of concentric rings around an unspecified globe (the Cakravala system, Buddhist inspired, in the sixth century AD), a disk floating on the back of a fish (Ainu belief in northern Japan), one of a number of flat platforms, which also included the underworld, connected by staircases (intermittent Babylonian philosophy), a labyrinth (a widespread, cross-cultural schema, including parts of India), the shape of tree (Scandinavian mythology) or as a globe (Greek philosophy of Anaximander (approximately 610–546 BC)). In all cases, the philosophical contemplations were unaffected by any direct observation of the earth; but such world-views fulfilled their purpose in explaining and propagating particular myths, in encouraging further investigation into the nature of the environment and in promoting a search for order and understanding.

Greek philosophy 550 BC to AD 150

The spherical nature of the earth was readily accepted by Greek philosophers by 550 BC. The pre-eminent school was that of Pythagoras (active in 530 BC), for whom the geometric perfection of the sphere was sufficient proof of its suitability as a framework for explaining the earth. The bounded and regular nature of the globe led to an over-optimistic opinion by such scholars that their current description of the earth was complete, or at least predictable. As Greek cartography moved towards a greater reliance on observation and measurement, shortcomings became apparent. Herodotus (489–425 BC) was among the first to propose a more empirical approach, relying on exploration and travel instead of pure geometry (in its modern sense) alone. Figure 1.1 shows the world map (about 500 BC) of his immediate predecessor, a major Greek historical and geographical writer, Hecataeus.

Figure 1.1   Map of Hecataeus (about 500 BC).

Map use in Ancient Greece was widespread and many episodes in Greek literature describe the everyday application of maps in law, travel and military campaigns. Despite the theoretical nature of much mapping, the educated populace was familiar with such documents. Later philosophers, e.g. Aristotle (384–322 BC), successfully married the concepts of Pythagoras with the pragmatism of Herodotus. The spherical nature of the earth was confirmed through observation of eclipses (the circular shadow of the earth on the moon, for example), recognition of the circumpolar motion of the stars and knowledge that the (northern hemisphere) Pole Star changed its angle of elevation with change in latitude. Information regarding the perceived inhabitable and inhospitable zones of the earth was incorporated by Aristotle, along with his description of a 'system of winds', to give a philosophically coherent and balanced world-view, sufficient to serve the Eastern Mediterranean culture of the period from the sixth century BC to the culmination of Greek cartography with Ptolemy (see Box 1.2) during the second century AD.

Religion and its influence on mapping

Well before the philosophical musings of early civilizations, which were based on the rationality and logic of those times, early societies invoked superstition and supernatural beliefs in attempting to explain the nature of the earth and expound a world-view. The development of organized religion was, according to one's opinion, a logical progression of such behaviour or an enlightened response to its deviance. Religious belief has had an enormous impact on the development of civilizations throughout all parts of the earth and everyday activity in many societies has been, and is still today, governed by behavioural codes

Box 1.2   Personality box – Ptolemy and the scientific nature of Greek cartography

Claudius Ptolemy (approximately AD 90–168) was unique in playing a role as the distinctive cartographic thinker and practitioner of two markedly different periods in European history, separated by over 1300 years. In the interim his writings were of value to map-makers of other cultures, notably Islamic and Byzantine civilizations east of Europe. His life's work, emanating from his home in Alexandria (site of the pre-eminent library of the age) in the second century AD, is considered to be the culmination of Greek map-making practice, and for a period in the fifteenth century AD translations of his work, from Byzantine copies (in Greek) into the Latin idiom, were regarded in western Mediterranean and north European societies as the complete source of geographical knowledge.

Ptolemy's writings on cartography cover instructions on map-making, discussions on the nature of map scale, mathematical descriptions of map projections and gazetteers of known geographical features. It is debatable whether Ptolemy himself produced the complete compendium of his cartographic work, the eight-volume work *Geographia*, but it forms the summary of his contribution to map-making. *Geographia* started with comments on the cartography of the day, personified by his near-contemporary, Marinus of Tyre. Marinus' cartography was criticized by Ptolemy: by not making good use of astronomical observation Marinus had overestimated the size of the inhabited world (*oecumene* in Greek); his descriptions of map projections were faulty (Marinus' proposed map of the *oecumene*, covering about one-quarter of the globe, was based on the false presumption that all the parallels of latitude in the area had the same length); and the content of the geographical data varied, seemingly randomly, from one edition to another. Written explicitly as a manual for map-makers, *Geographia* included an 8000 entry gazetteer and a description of 27 regional maps which were to supplement the text. Ptolemy proposed three map projections in great detail: similar in contemporary terms to an equidistant conical; an approximately equal-area pseudo-conical, with curved parallels and meridians emphasizing the spherical nature of the earth; and a perspective azimuthal. These notes form the most lasting Ptolemaic legacy.

Ptolemy reflected the Graeco-Roman interpretation of cartography and it is ironic that *Geographia* first appeared to European scholars in 1406, almost at the end of the dominance of the Roman-inspired *mappae mundi* framework in cartography (see p. 15). The first printed edition of the translation appeared in 1475; the 1477 Bologna edition and further editions included the 27 reconstructed maps from Ptolemy's writings (these are sometimes called *tabulae modernae*). By the turn of the sixteenth century, in the light of contemporary knowledge, the shortcomings of the Ptolemaic world- and regional-views became apparent and newly created maps were included. It has been suggested that the increasing numbers of such new maps, bound together in book form, initiated the development of atlas cartography.

Figure 1.2   The world-view of the Roman Empire (c. AD 400).

and standards directly obtained from sacred teachings. Mapping, embracing as it does a set of universal truths (the shape of the earth, the arrangement of the cosmos, the order imposed by distance and direction), has often been thought to be strongly influenced by doctrinal rigour and dominant belief systems.

In fact, cartography has proved to be more stubborn than many fields of human activity in resisting the clamour of prevalent dogmas and the narrow-mindedness of some periods of religious intolerance. Maps themselves, unlike other creative artefacts such as representational art, icons, sculpture and manu-scripts, have never been considered as sacred items in their own right (with a very few early Chinese and Egyptian exceptions). There have been few attempts through history, therefore, to be prescriptive of the nature, content and appearance of maps. Even when changing world-views have been devel-oped and been resisted by the religious establishment (e.g. the thoughts of Copernicus published in AD 1543) the impact on cartography has generally been minimal.

However, it should be acknowledged that there have been significant periods in the history of cartography in many societies where the accepted world-view (religious beliefs have had little impact on large-scale, local mapping at any time) has been influenced to a greater or lesser extent by religious belief, or, more likely, by restriction on new thinking due to the impact of religion on current intel-lectual endeavour as a whole.

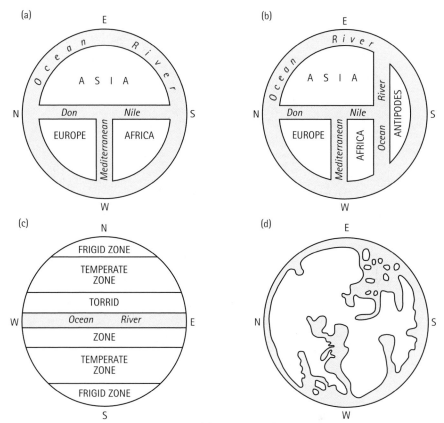

Figure 1.3 Typology of *mappae mundi*: (a) tripartite; (b) quadripartite; (c) zonal; (d) transitional (from Harley and Woodward, 1987).

Mappae mundi and European cartography AD 400 to AD 1450

One such period, often derisively dismissed as a period of 'church maps', was the thousand year era that encompassed the 'Dark' and 'Medieval' Ages in Europe. During this time the intellectual legacy of Greek and Roman civilizations was predominant but notably subdued. Despite some later Islamic influence, there was, in general, little inflow of new cultural ideas from elsewhere. In terms of the world-view, the prevailing framework, itself a derivation from the cartographic products of Ancient Greece, was that furnished by the decaying Roman Empire (Figure 1.2). Thus, a circular earth disc, set in a surrounding ocean, became the dominant interpretation of the Middle Ages cartographer. Often modified and varying in appearance through time (see below), the maps that were produced have collectively been given the Latin name *mappae mundi*, 'maps of the world'. Alternatively, the nature of their layout has given rise to the term 'T in O' maps, in which the Mediterranean Sea forms the stem of the letter 'T', the rivers Don and Nile form the horizontal bar, and the land areas of Europe, Asia and Africa assume the 'O' shape (Figure 1.3).

Like the Roman world maps, the prime purpose of the *mappae mundi* was

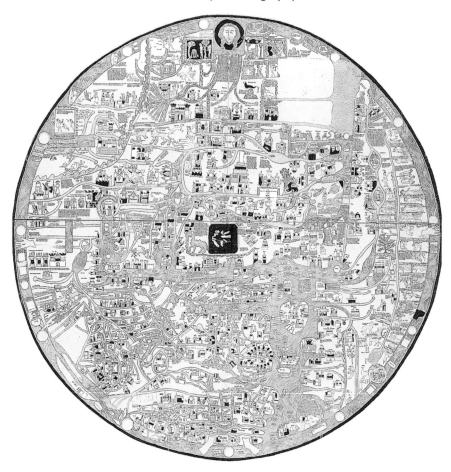

Figure 1.4 The Ebstorf map showing Christ's head, hands and feet at the extents of the world (AD 1235).

not to show location. With little international trade, insignificant cultural links, minimal diplomatic contact and few opportunities or needs to be aware of distant locations, there was little 'map use' to speak of. The major role of some *mappae mundi* was to relate a narrative, usually with an historical theme, acting almost as a medieval cartoon in a similar fashion to stained-glass windows in the great European cathedrals of the time. The essence of the story was to demonstrate the dominion of Christ over the face of the earth (Figure 1.4) or to illustrate certain Old Testament Biblical tales. A large number of *mappae mundi*, however, exhibit little overt Biblical content (often depicting other legends and fables instead) and although the geographic framework is irregular and the content fanciful, such map products are not quite as rudimentary as some historians of cartography would suggest.

The epoch of *mappae mundi* predominance in European cartography can be divided into four gradually merging periods. The time immediately after the collapse of the Roman Empire, from AD 400 to AD 700 was, in western Europe, an era of limited intellectual opportunity and consequently little cartographic output. The inherited Roman framework of the world was little modified: the division of the earth into three parts was made to correspond to the peopling of the earth by the descendants of the three sons of Noah; the lack of a graticule was probably not even considered important; and the symmetry of the Mediterranean, Don and Nile axes allowed orientation to the east with Jerusalem locatable at the centre of the map (although such positioning is, in fact, not the case for the majority of *mappae mundi*). Scholarly output was considerably more developed from AD 700 to AD 1100, emanating mainly from monastic institutions. Religious influences, therefore, were growing. The maps of the Venerable Bede (AD 672–735) from northern England and Beatus of Liebana (active AD 776–786) from Spain, for example, openly show the Apostles proselytizing into all parts of the world. The period of the Crusades and the initiation of mass pilgrimage (AD 1100 to AD 1300) called for map documents which depicted the Holy Land in particular. The maps from this era are less schematic than previously but are often supplemented by more overt religious instruction in the form of sketches showing the stages of life and of pilgrimage from a Christian perspective. The transition period between the well-developed 'T in O' maps and an age of cartography more dominated by secular factors – such as the re-emergence of the Ptolemaic coordinate system and the impact of the accurate navigational charts of the Mediterranean, called portolans – was from AD 1300 to AD 1450.

To an extent, the religious establishment throughout the whole period was indifferent to map-making. Knowledge of the physical world was considered secondary to the required knowledge of the spiritual realm: maps were unimportant artefacts with little direct use beyond decoration. Although Biblical influences did increase over the thousand year period as the classical inheritance of Greece and Rome became ever more distant, there was a simultaneous increase in new information, particularly in the final stage described above. The output of serious scholars and cartographers, such as Fra Mauro, whose most famous cartographic work dates from 1459, showed a significant breakaway from what remaining influence religion had on map-making by the fifteenth century.

Travel and its influence on mapping

The European 'Dark Ages' was one of the few periods in history when the human tendency to be a mobile species was severely curtailed, at least in that part of the world. Humans have been called the most restless creatures on the earth's surface, their desire to travel being prompted by a variety of activities including foraging, exploration, warfare, socialization, tourism, commerce, pilgrimage and diplomacy. In each case the need for maps has usually been evident and, in addition, the contribution of such travels to the total amount of human spatial knowledge has been clear. The role of the traveller in everyday map-making has often been overlooked in histories of cartography. The world-views propounded by the philosphers of differing cultures were more varied and have held more intrinsic appeal, and, in general, they have survived longer as artefacts: travel maps naturally become worn out and are discarded much more readily. The frequent disparity between the map of the philosopher and the map of the traveller has been repeated in numerous societies throughout the ages. The map, in the hands of the former, was symbolic; while on the ship or in the camel train of the latter it was a working instrument.

The 'Age of Discovery'

From a European perspective, the Age of Discovery (starting approximately AD 1450) describes the period of initial colonization of distant parts of the globe. Despite the impact of the Crusades on the mobility of western and northern Europeans, eastwards movement beyond the Mediterranean was, in general, restricted for several centuries after AD 1280 due to the strength of the Ottoman Empire. Burgeoning travel, fuelled by commercial and colonial impulses, was therefore destined to be westwards. The Portuguese were the main initiators of such expeditions. During the middle of the fifteenth century, the western coast of Africa became familiar to Portuguese navigators, and by 1497, a route around the Cape of Good Hope towards the suspected riches of Asia had been opened up by Vasco da Gama. Portuguese charts showed a remarkable level of detail of the East African coast, India and islands such as Madagascar and current-day Sri Lanka.

Much of the impetus for voyages towards the East had come from accounts of the societies encountered by travellers such as Marco Polo who had earlier (in lengthy journeys from 1271 to 1295) purportedly spent significant periods in China. Marco Polo's information, much of it (such as knowledge of the existence of Japan) repeated from Chinese material, did not have much impact on the still dominant *mappae mundi* of the time, although his observations did prompt immense interest in subsequent centuries. The Fra Mauro map shows evidence of the inclusion of information from this source.

Portuguese navigational achievements were initiated by expansionist and colonial policies and prompted other European nations also to explore and colonize. The period from 1490 to 1510, in particular, saw a large increase in European knowledge about the coasts of Africa, Asia and the Americas. The Spaniards, in expeditions such as those led by Columbus, explored the Caribbean Sea, its shores and islands; the English expeditions under John Cabot made landfall farther north in Newfoundland; and the Portuguese began their colonization of the coastline of present-day Brazil. As early as 1500, detailed charts showed much of the coastline of North and South America and the earliest map showing the Americas as a separate land mass was drawn by Martin Waldseemuller in 1507. In addition, although the Dutch are credited with the first encounter in 1606, it is suggested that the Portuguese, considerably earlier, were the first Europeans to arrive in Australia, leaving its shore off their charts for reasons of secrecy and fear of rival European nations.

Most of the maps produced for and by such seaborne expeditions were, naturally, navigational charts for sailing ships. The portolan charts had been gradually increasing in importance as trade and travel in the Mediterranean Sea developed from the thirteenth century onwards. The portolan charts, as practical documents, were significantly different in appearance and function from the contemporary *mappae mundi*. They were created using overwhelming reliance on nautical observation and their level of planimetric accuracy was high. Distances and bearing between all the trading ports and safe havens were well known, and, using these, an iterative construction process led to the establishment of a firm framework of recognized points that could be accurately plotted by the cartographer of any portolan chart. The main centres of such production were in Italy, notably Venice, and in Catalonia. Apart from the locations of harbours and the shape of the coastline, there was little detail on these charts. Landward detail was minimal, being of little relevance to the seaborne navigator, and the land areas provided space for compass roses, scale bars and written notes. The sea area was similarly blank to allow for the plotting of routes. The whole surface of such charts was covered by regular arrangements of rhumb lines which, it is suspected, were used to help in accurate navigation. Notable research in the history of cartography has resulted from the analysis of the portrayal of geographic names ('toponyms') on such charts and by the examination of the flags, heraldic devices and symbols which were generally placed next to the prominent cities of the Mediterranean region. Such clues can give insights into the contemporary political geography of the area from the thirteenth to the fifteenth centuries. Later Catalonian portolan charts tended to include significantly more land detail, including rivers and mountain ranges. In effect, they became more general purpose regional maps.

Those whose task is navigation (see pp. 45–53) have always been prime consumers of cartographic products. In the route-planning stage, during the journey itself, and whilst reflecting later on the trip, maps have shown their utility. The high-accuracy navigation of a supertanker captain at sea, a military pilot on manoeuvre or an orienteer running in the forest requires map products that are developments of early map-making. More general products such as maps for tourists, for the leisure motorist and for the business traveller are produced by the million every year. Travel, in all its guises, has an intimate need for cartographic output.

Science and technology and their influence on mapping

Developments in science have affected the perceived nature of the framework within which mapping has taken place (i.e. the shape and dimensions of the planet), the subject-matter and content of the map and the method by which the map itself has been manufactured. The impact of scientific discovery has, however, been erratic and diverse. Elementary surveying instruments such as the spirit level and the compass were known in China (Asia's oldest civilization with a primacy in cartography) before 1100 BC, yet not developed in Europe until at least a millennium later. Further developments, including paper manufacture, advances in astronomical observation and woodblock printing ensured that

Chinese cartography of the period AD 400–1450 was radically different from the European picture described on p. 13. Amongst the wide range of maps created during this period, Chinese cartography was characterized by a certain uniformity, in scale, appearance and function, based on the fundamental framework provided by accurate scientific observation and measurement and by the printing technology available to ensure identical multiple reproductions of maps.

The 'reformation' of cartography: eighteenth-century France

In Europe, particularly in the Low Countries, the evolution of similar technologies, most notably printing, had an impact on the extraordinary flowering of cartographic activity in the first half of the seventeenth century. However, it was not until later in that century and into the eighteenth century that a truly 'scientific' climate matured in western Europe. The most fertile ground for such innovation was France which, from 1660, was a centralized and bureaucratic state under the newly installed monarch, Louis XIV. Despite lagging behind the Netherlands and England in the techniques of map-making, navigation, applied mathematics and instrument-making, French investigation into the more fundamental aspects, most notably geodesy, triggered growth in these associated areas also.

Early seventeenth-century mapping, especially at small scales, but even for large-scale maps used as working documents, had been highly decorative with considerable embellishment on the map face in the form of heraldic devices such as coats of arms, sketches and portraits, swirling and detailed title boxes and pictures of heavenly cherubs. The reformation in cartography, however, supplanted this pictorial tradition. The new cartography in France, introduced by its first official cartographer, Nicolas Sanson (1600–1667), also relied less on scholarly compilation of written descriptions and more on instruments and measurement. A bureaucratic state needed information on its true size and shape, as well as an indication of its socio-economic circumstances. These required an observation-based, scientific approach to data collection. This was achieved using established field-surveying techniques, particularly plane tabling; but newer methods relying on accurate observation of both horizontal position ('planimetry') and of height with theodolites, levels and improved telescopes led to the refinement of 'triangulation', providing an accurate framework of known points throughout a widespread area. Improved mathematical methods (using newly created logarithms and trigonometrical tables) allowed for more sophisticated manipulation of the field data. The scientific approach also affected thematic mapping, particularly later towards the end of the eighteenth century. Considerably more maps were being made of measurable phenomena such as economic production, agricultural distributions, demography, traffic flow and educational level, and maps were made to support statistical analysis of such geographical features. However, cartography had to wait until the nineteenth century for the development of adequate symbolizations and designs to convey these ideas efficiently.

The prime movers in the mapping of France and ensuring its adherence to

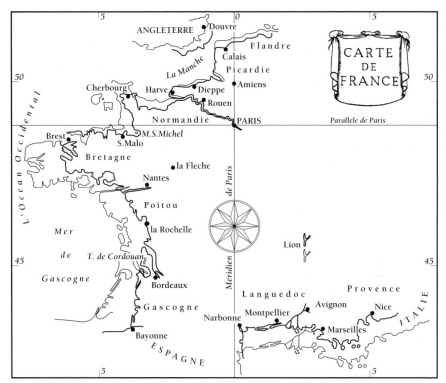

Figure 1.5 The perceived shape of France before and after the 1693 survey by Picard and La Hire (revised coastline in bold).

scientific fundamentals were successive generations of the Cassini family, naturalized Italians attached to the French court throughout the eighteenth century. Geodetic techniques were employed by them, and others, to determine the outline of the country and form the basis of the first uniform topographic map series of a complete nation-state. Cassini I (1625–1712) initiated measurements along the meridian of Paris using triangulation techniques in 1669. Regular checking of latitude position using established methods of field astronomy allowed, for the first time, accurate estimations of eastings and northings on land. Further, accurately timed astronomical observations helped determine the positioning of the meridian passing through the Paris Observatory; and further triangulation ensured its accurate location, by 1718, across France, south to north, from the Pyrenees to the English Channel.

Triangulation was also undertaken along the entire French coast by Picard and La Hire in 1693. The result drastically re-appraised the shape and size of French territory, slimming it down by 1.5° of longitude and 0.5° of latitude (Figure 1.5). The triangulation scheme was then extended over the whole of France and a new systematic map of France based on this accurate framework was to be prepared (the idea had been put forward by Cassini I in 1681). The survey work was undertaken between 1733 and 1783, and was carried out by Cassini II (1677–1756) and Cassini III (1714–1784).

The 'scientific' map-making of France based on the work of the Cassini family was started in 1747 and was initially subsidized by the state. The support was withdrawn nine years later but Cassini III carried on at his own expense. By the time he died only Brittany remained to be completed. After a short suspension during the French Revolution, the whole project was finished in 1818. It consisted of 182 sheets at 1:86 400 scale (1″ to 2400 yards), engraved on copper plates. All the large towns were shown in plan view (up to this time most three-dimensional structures such as hills and buildings had been shown in profile); a variety of symbols were used to mark churches, mills and even gallows. Forests were portrayed as well as paths within them. Further innovation occurred with the use of hachures, rather than perspective hill drawing, for relief representation. This was the most advanced and innovative cartographic product of its time. It relied on accurate scientific measurement in the field, and also on improved cartographic design and representation.

The French were interested in mapping activity beyond their shores – they had colonial interests in West Africa, in the Caribbean and in southern Asia. Nautical charting was undertaken in these areas, but the most important French overseas work was in geodetic observation. The work done within France itself had led to enormous advances in geodesy and interest in global applications of the new sciences of measurement. Despite the predominance of a Newtonian view of the world (first published by Newton in 1687 and widely accepted by 1730) which argued the opposite, the Cassinis were convinced, from their limited measurements in the mid-latitudes of northern Europe, that the shape of the earth was elongated at the poles. French-supported expeditions were sent towards the North Pole (through Lapland) and towards the equator (in current-day Peru) in the 1740s and confirmation of the Newtonian prediction of the earth as being flattened at the poles was obtained. A member of the Peruvian expedition, Pierre Bouguer, performed valuable analysis of gravity, noting the influence of large rock masses (i.e. mountain ranges) and of height above sea-level on geodetic measurements. Further scientific investigation included the linking of the Greenwich and Paris observatories, and independent observations led to precise estimates for the width of the English Channel (further described in Box 5.4).

The scientific approach to cartography was part of a wider change in the scientific–societal context which affected social, political and economic issues. This manifested itself in map output in the form of cadastral mapping: land registration was needed for fiscal and juridicial purposes, and wider map accuracy was also sought to allow for fairness in the determination of government representation in post-revolutionary France. The perceived importance of *égalité* was instrumental in promoting the implementation of accurate, 'scientific' map-making.

The world-wide impact of changes in science and technology has not abated since the time of the Cassinis. Particular periods of accelerated development can be noted, e.g. the diffusion of lithographic colour printing in national mapping agencies and its use by atlas publishers in the 1880s; the widespread adoption of photogrammetry for topographic mapping in the 1940s; and the introduction of plastic drafting and reproduction materials in the 1950s. However, most of these periods have exhibited progress only in certain elements of the map-making flowline, such as data collection or map reproduction. Currently, in the 1990s, as exemplified on p. 77, we are in the midst of a scientific revolution in *all* aspects of mapping – data collection, data storage, data manipulation, data presentation and map use and interpretation.

The Islamic tradition in map-making

One purpose of this chapter is to demonstrate that the level of map-making activity is a function of the society and culture within which it is undertaken. As has been shown, most societies, both concurrently and sequentially throughout history, have relied upon each other for the development of such activity. However, individual societies have exhibited considerable variation in their level of map-making achievement and its evolution. This section considers one example, showing the nature of map-making in the context of Islamic culture and the societies within which such a culture has been the primary determinant of human activity.

The earlier decline of Greek influence in the eastern Mediterranean and the waning of the Roman Empire enhanced the impact of the newly formed Islamic culture in the whole of the Mediterranean Basin and the Middle East from the seventh century AD onward. As with the societies and map-making undertakings highlighted in the sections above, the history of pre-modern Islamic civilization and its cartographic production, is varied. This reflects the differing scientific and technological influences through time, from both west (primarily Greek tradition) and east (Indian subcontinent influence); the range of travel undertaken by Arab and Persian merchants and explorers, from Siberia to Morocco and from Indonesia to Finland; and the changing philosophical balance between theoretical and practical approaches to map-making. In addition, Islamic cartography shares with other cartographies factors such as religious and political influence from the patronage of wealthy and powerful individuals; significant inheritance from previous traditions and the establishment of values and continuity for succeeding generations; and an incorporation of cartography with other 'communication' methods, such as written texts and pictures to create 'knowledge stores', which summarize the wisdom of a civilization for posterity. The influence of textual material in Islamic geography is particularly notable: most of the maps now extant were supplements to, or incorporated within, written manuscripts.

Three distinct phases of Islamic cartography can be described, although there is significant overlap in their chronological sequence. The transitional phases occurred as a result of many internal and external influences, e.g. the clear geographical requirements for fulfilling religious observance, the varying political and economic impact of western civilizations and the maturing of Islamic scientific activity in academies of scholars. The earliest stage, starting in the eighth and ninth centuries AD, was primarily influenced by the Greek scientific tradition and practical map-making skills. Considerable efforts were made in

Baghdad, the capital of the Abbasid Empire, to translate into Arabic (a relatively new language) scientific works from a range of neighbouring societies, east and west. Knowledge of the earth was summarized in tables by scholars such as Al-Farghani (active in AD 861). His tables, both geographical and astronomical, were not precise enough to allow for map production and they are thus dissimilar to the gazetteers of Ptolemy. Other contemporaries, however, such as Al-Khwarazmi (died AD 847) and Al-Battani (died AD 929) did have access to translations of Ptolemaic writing and copies of his maps, from which, it is suggested, the former merely copied coordinate information: much of the material produced during this period was highly derivative. Ptolemy's *Geographia* was translated along with his astronomy-based *Syntaxis* (translated into Arabic as the *Almagest*) which was regarded considerably more favourably by Arab scholars, although it is suggested that there was little interchange of data between astronomers and geographers.

The second phase of Islamic cartography, which was considerably more independent, was exemplified by Al-Balkhi (died AD 934), who gave his name to a distinctive school of geographical thought. This concentrated on regional mapping, particularly the *aqalim* (regions) of Persia, supplemented by written notes. A few world maps were produced, but many of these were becoming highly stylized, with regular geometrical shapes providing the basis for the outlines of continents. Some of these maps became as symbolic as the 'T in O' maps were for western European societies of that time. However, the breadth of Islamic influence and its sway over a number of diverse cultures ensured that there would still remain some intellectuals concerned with a more empirical cartography. Al-Biruni (AD 973–1050) was one such scholar, well versed in both Greek and Indian scientific traditions. His writings concentrated on the determination of the *qibla* (the sacred directions for Mecca), but he also carried out significant experiments to measure the length of angular distances (e.g. one degree of longitude, which varies in length at differing latitudes) in various parts of the earth; he proposed new map projections; created more comprehensive geographical tables; and mused, philosophically, on the distribution of earth and water on the planet's surface.

As westward progress of Arab civilization was denied by European states during this middle phase of Islamic cartography, whatever external influences there were came from the East and also from northward journeys, particularly along the Volga River. Resultant mapping reflected this, the maps of Al-Qazwini (AD 1281–1349), for example, incorporating Chinese influence in the form of a rectilinear grid system and including the Mongol names for Asian flora and fauna depicted on the map.

The final stage of Islamic cartography returned to the influence of western ideas, an influence that was not reciprocated. The work of Idrisi (AD 1100–1170) was pre-eminent during this phase. Born in Morocco and educated in Spain (at Cordova), Idrisi was employed at the court of King Roger I in Sicily, like him a Norman-Arab. The court in Palermo was in a useful location to receive ideas from many different societies and Roger I was keen to support scientific investigation and geographical expeditions. Idrisi's main output (AD

1154) consisted of a major geographical treatise in manuscript, 70 sheets of regional mapping and a world map. The latitude-based framework of habitable zones, introduced by the Greeks, was adapted by Idrisi who produced maps with climatic areas called *kishvars* (Persian = Arabic *aqalim*), more numerous and extensive than the Greek schema.

The later history of Islamic cartography is diverse: the Balkhi school continued cartographic output into the eighteenth century; Idrisi's influence was maintained, particularly in North Africa, for a significant period up to the seventeenth century; and further east, the influence of Portolan charts was being felt on both Islamic cartographers and those of other civilizations such as the Ottoman Empire, both of which started producing accurate navigational charts.

Summary

The case study just described encapsulates many of the determining factors that have influenced the history of cartography. Their impact has been highly variable, in time and in space, and, as was suggested at the beginning of this chapter, there is little sense in trying to generalize about the development of map-making activity. However, all the factors evident in the Islamic tradition – the effects of 'international' (or at least inter-societal) contacts, the role of 'world-views' and views of the cosmos, the impact of male-dominated map-making activity (the influence of gender on map-making is discussed further on p. 93), the influences of philosophy, religion, science and travel, and of everyday human activity on the soil or in the town – have likewise contributed to the overall picture of cartographic development world-wide.

It is clear that 'objective' scientific factors have been, and are, merely one strand (albeit important) in the story of mapping and map-making. The suggestion that map-making 'progresses' through a concentration on the perceived accuracy (as scientifically measured) of the map does not take into account the range of other strands which contribute to map production. This chapter has considered some of these – philosophy, religion and travel – and further factors, such as societal *praxis*, economic circumstances and political decision-making, will be discussed throughout the rest of the book in their contemporary setting.

It is important to note that the history of cartography is not 'finished'. Although most creative and intellectual effort by historians has been directed towards various perceived 'golden ages' of cartography or towards the products of long-since disappeared civilizations, current cartographic practice is, of course, merely one further stage in the course of events that have unfolded over many centuries, and is eminently worthy of study in its historical context. The remainder of the book will address, primarily, contemporary issues which affect mapping and map-making, but these are all built on historical foundations and will themselves form the basis for description and analysis by future historians. Among these historical foundations are factors common to all maps – scale, map projection and depiction of spatial data: these are considered in the next chapter.

Further reading

The benchmark for writing on the history of cartography is the lavish set of volumes edited by J. B. Harley and D. Woodward, *The History of Cartography* (University of Chicago Press, Chicago). This multi-volume, multi-author work is an extraordinary and ongoing publishing achievement, encyclopaedic in its coverage, unrivalled in its depth and profusely illustrated. Volume 1, *Cartography in Prehistoric, Ancient and Medieval Europe and the Mediterranean* (1987), Volume 2 (Book 1), *Cartography in the Traditional Islamic and South Asian Societies* (1992) and Volume 2 (Book 2), *Cartography in Traditional East and Southeast Asian Societies* (1994) have been published. Further intended volumes are Volume 2 (Book 3), *Cartography in Traditional African, American, Arctic, Australian and Pacific Societies;* Volume 3, *The European Renaissance* (which will include accounts of the golden era of cartographic dominance by the Low Countries); Volume 4, *The European Enlightenment;* Volume 5, *The Nineteenth Century;* and Volume 6, *The Twentieth Century.* Background reading preparatory to studying the history of cartography should include M. Blakemore and J. B. Harley, *Concepts in the History of Cartography*, Cartographica Monograph 26 (University of Toronto Press, Toronto, 1980). Comprehensive overviews of the history of cartography which generally take a chronological perspective include the early classical work by L. Bagrow (revised and edited in translation by R. A. Skelton), *The History of Cartography* (Precedent, Chicago, 1964) and the more modern J. Goss, *The Map Maker's Art* (Studio Press, London, 1993). *A History of Cartography* by C. Bricker (Thames and Hudson, London, 1969) takes a novel continent-by-continent perspective. The role of Ptolemy is given prominence in Chapter 11 of *The History of Cartography*, Volume 1, by O. Dilke. Portolan charts are detailed in Chapter 19 of *The History of Cartography*, Volume 1, by T. Campbell. An account of the reformation of cartography in France which admirably sets map-making in its societal context can be obtained from J. Konvitz, *Cartography in France 1660–1848* (University of Chicago Press, Chicago, 1987). Islamic cartography is covered in great detail in Chapters 1–9 of *The History of Cartography*, Volume 2 (Book 1).

Chapter 2

The shape and content of maps

Maps and their scale

The history of cartography reveals the continuity of people's attempts to bring order to their 'reality' by setting down their world-view in map form. Map products vary enormously, but they all share the distinction of being spatial descriptions of the world or part of it. Fundamental to such representations are a reliance on *scale* to bring the world-view to manageable proportions, a consideration of *map projection* to ensure that the irregularity of the earth's surface can be precisely addressed on a two-dimensional plane, and an appreciation of the relevant *content* to portray. This chapter addresses each of these issues.

Some maps represent only small portions of the earth's surface; they portray the landscape, other spatial features and their variation in great detail over a limited tract of space. Such maps are *large-scale* maps, because the *representative fraction,* which indicates their scale, is relatively large. Thus, a plan of a local urban park at a scale of 1/500 (1 unit of measurement, e.g. 1 cm, on the map is equivalent to 500 such units of measurement (5 m) on the ground) has a large scale. By contrast, a map representing the entire planet at a scale with representative fraction 1/25 000 000 (1 cm is equivalent to 250 km on the ground) is a *small-scale* map. The boundary between large- and small-scale maps is subject to enormous subjective individual variation such that it cannot be precisely defined. This is of little concern as the terms are comparative, not precise.

However, maps that represent the entire earth on one sheet of paper or on one computer screen are certainly small-scale maps. Such images are ubiquitous. The world map has become a motif for organizations ranging from the United Nations to the CNN media network. It is enshrined on the wall of the classroom as an icon of knowledge and it is often presented to students of geography as the symbol which binds the discipline together – brandished on the cover of numerous textbooks, including this one. Outside the classroom it is placed behind the newscaster's head, on the T-shirt and on the box of eco-friendly washing powder: it is the most widely used image of the 1990s (see Box 2.1).

The graticule

In its most common form the world map consists of the outline of the continents, floating on an indistinguishable background – the oceans which surround us, but of which we know less than we do about the surface of the moon. The

Box 2.1   Contemporary mapping box – The map icon in entertainment, communication and advertising

For the majority of the world's population, the most familiar map is one representing the entire globe. This fact is recognized by many of the multinational corporations which impact or endeavour to have an effect on people's everyday lives. The malleability of the world outline, its familiarity to most cultures, and the desire that many contemporary entertainment, communication and advertising industries, for example, have for presenting themselves globally and reflecting their world-wide interests, have ensured that use of the world image grows apace. Such images have, over the last few years particularly, become icons, and, in this context, are intended to depict and reflect the supposed powerful internationalism, environmental awareness and caring nature of these businesses. As with other icons in other ages, any questioning of these map images, their appearance, purpose and their subconscious message, is regarded as near-heretical. This lack of scepticism is often extended to all other maps, whose authority and veracity is unquestioned by most. Both small-scale world

(a)   (b)

(c)   (d)

Rendering the shape of Australia using (a) arrows (recycling logo); (b) stylized hands (land resource management agency); (c) text (national camp site business); (d) line symbols echoing extracts from the national flag (commemorative organization).

(continued)

(continued)

maps and large-scale official representations are regarded as 'truthful' interpretations of the space we inhabit. It will be shown later in this book that such universal acceptance of the world-view seen through the filter of a map may be misguided. However, for large multinational businesses, the global news corporation or the international aid agency, the world map icon is effective as an indicator to their customers, employees, viewers and donors of their worldwide status, concern and influence.

One reason for the widespread use of the world image is the instant recognition afforded to the outline of continents and the graticule, even when projected and distorted onto a flat sheet or screen. This universal familiarity with the shape of the earth and its land masses, extends also to the geographical outline of nations and regions. The figure shows the outline of Australia created by a variety of graphical devices. The range of institutions using such devices, exemplified here, is testament to the power of the map to symbolize the fundamental nature and scope of their business.

Despite artistic embellishment, the graphic designer, cartographer, cartoonist or propagandist has, in general, the confidence to assume that the viewer has such familiarity with a national outline that no caption is necessary: it can be represented by objects such as beer cans or agricultural produce, by its stretching to fit a set of alphabetical characters or by matching its shape to a human or animal caricature.

world map is capable of considerable distortion yet it still keeps its recognizability (Box 2.1). Its complete outline can be moulded into many shapes: rectangular, circular, oval, cordiform (heart-shaped), star-shaped and numerous others. The basic net of lines of latitude (parallels) and lines of longitude (meridians) is called the *graticule* and it is onto this fundamental framework that the shapes of the continents are placed. The graticule is usually plotted as a regular series of connected *x,y* Cartesian coordinate points (or, less often, in a polar coordinate system, *r,*ϕ), positioned according to the *projection* used.

The rectangular representation of the earth is probably the best known, due to its renditioning in a number of widely used map projections, and because it matches the shape of the atlas page or the computer screen. In the past, it has proved a beneficial shape for those presenting a particular world-view. Usually centred on the Greenwich meridian (0° of longitude) and often showing the northern hemisphere (with its greater land-mass area) as greater in proportional area to the southern hemisphere, the rectangular world map has a focal point in Europe, placed as surely at the centre of the world as some European medieval maps placed Jerusalem. More recently, a North American perspective can be taken to justify such maps which have the United States of America in the 'top left' (the traditional place in western culture to start reading a document), imperiously watching over the rest of the world.

In fact, American cartographers have been in the vanguard of attempts to resist the use of such rectangular map projections for the representation of world-views. These maps show the earth as having straight edges and sharp corners (rendering each parallel and meridian as straight lines when in reality each of them is circular), they represent most distances and routes incorrectly, and they show the lines of latitude as equal in length when in fact they are all shorter than the equator. These disadvantages are seen as particularly important to avoid when designing maps for schoolchildren or an unsophisticated audience, and some cartographers have been spreading the message that they can be overcome by using more appropriate world map projections (see Further reading).

Map projections

Whatever projection is chosen, however, the graticule and world outline have, out of cartographic necessity, to submit to some form of distortion. It is impossible to represent the intricate and wrinkled surface of our globular planet on a flat plane, paper sheet or monitor screen, without some form of deformation being introduced. At the small scales at which the entire earth can be mapped as one image, the requirement is to 'project' the spherical representation of the earth onto a flat surface. Map projections, of which there are many, each with its own merits and drawbacks, are the mechanism by which such transformations can be undertaken. It is important to realize that all map projections are in fact mathematical transformations which create a new set of x,y (or r,ϕ) points defining the shape of the graticule and positions on the sphere, from the real-world points defined by latitude and longitude. The values of x and y are therefore determined by the nominal scale of the map that is required, the latitude and the longitude of the location, and the mathematical function that is applied to them.

The creation of a map projection: an example

In order to create a map projection it is necessary to consider the mathematical functions that are applied to the real-world locations in order to obtain coordinates that can be plotted by hand or (more usually) by computer onto a sheet of paper or a monitor screen. The results of applying these functions are visual representations of the graticule and geographical locations on a flat surface whose properties (see next section) are known. Table 2.1 gives an indication of the results of the coordinate transformations for the globe undertaken when the sinusoidal (also known as the Sanson-Flamsteed) equal-area projection is calculated. For each point,

x = scale $\times$ longitude $\times$ cosine (latitude)
y = scale $\times$ latitude

In this example the nominal scale is 1:120 000 000.

The angular measurements have to be computed in radians and the scale is the representative fraction of the nominal scale applied to the radius of the earth (6371 km = 637 100 000 cm). The x,y coordinates in Table 2.1 (in centimetres)

Table 2.1 *x,y* plotting coordinates (in cm) for the north-east quadrant of the world sinusoidal projection (scale 1:120 000 000)

Latitude	\<Longitude\> 0°	30°	60°	90°	120°	150°	180°
0°	0, 0	2.78, 0	5.56, 0	8.34, 0	11.12, 0	13.90, 0	16.68, 0
20°	0, 1.85	2.61, 1.85	5.22, 1.85	7.84, 1.85	10.45, 1.85	13.06, 1.85	15.67, 1.85
40°	0, 3.71	2.13, 3.71	4.26, 3.71	6.39, 3.71	8.52, 3.71	10.61, 3.71	12.78, 3.71
60°	0, 5.56	1.39, 5.56	2.78, 5.56	4.17, 5.56	5.56, 5.56	6.95, 5.56	8.34, 5.56
80°	0, 7.41	0.48, 7.41	0.97, 7.41	1.45, 7.41	1.93, 7.41	2.41, 7.41	2.90, 7.41
90° (North Pole)	0, 8.34	0, 8.34	0, 8.34	0, 8.34	0, 8.34	0, 8.34	0, 8.34

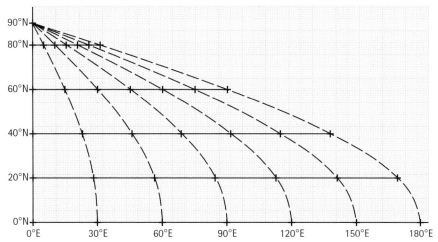

Figure 2.1 Plotted sinusoidal equal-area projection of part of the globe on a graph paper background with emphasized calculated graticule intersections from Table 2.1.

give the north-east quadrant, but the projection is symmetrical and positions in all the other quadrants can be easily deduced. The plotted result, showing each *x,y* point in the table correctly placed on a graph paper background, is shown in Figure 2.1.

For some locations, some transformations do not give calculable answers (e.g. the north and south poles cannot be shown on a Mercator projection); in other cases real-world *points* are transformed into *lines* on the map. It is the transformations which have the characteristic of bringing about deformations at some points on the map (sometimes at every point on the map). This means that the scale of a map is only valid at certain points or on certain lines on the map. The 'nominal' scale of a world map should, therefore, be treated with caution. Because distances are usually not scaled correctly or evenly across the map, other properties which one might wish to measure from a map and which can be defined with reference to distances, such as the bearing of one point from another, or the surface area of a region, are also subject to the inherent distortion. However, although no flat projection can represent all distances correctly, many can represent bearings or areas without distortion. The mathematical properties of projections are discussed below, followed by other considerations which can be taken into account when choosing a reference frame within which to present a world-view.

Map projection properties

Those projections which do represent distance scaled correctly can only do so in one direction (usually north–south, such that the parallels are correctly spaced across the meridians), although they usually have correct scale also along one line (the line of zero distortion) in the perpendicular direction, e.g. at the equator. Such projections are termed 'equidistant' projections. However, it should be noted that this does not mean that the distances are correctly scaled everywhere on the map. The world outline cannot be represented at true scale throughout except on a globe. Equidistant projections are often more aesthetically pleasing for representing large portions of the earth's surface than the alternatives discussed next.

A map on which bearings are rendered as they appear in the real world is represented with a 'conformal' projection. The Mercator projection is an example of a conformal map projection, and it has a primary use in navigation where correct representation of bearings and angular measurement are important. Because bearings are truly represented, shapes (over small areas) are preserved on a conformal map projection. Unfortunately, because scale is increasingly distorted away from the line or point of true scale, areas are exaggerated, sometimes grotesquely, and for maps which portray large areas of the earth's surface (small-scale maps), conformal mapping is not appropriate.

A map which portrays areas on the earth's surface in their true proportion uses an 'equal-area' or 'equivalent' projection. Such representations have application where the cartographer wishes to show distributions on the earth's surface which can be compared and contrasted, e.g. the relative areas of different vegetation types, or for certain forms of statistical mapping. Because an equal-area projection must compensate for the inevitable scale distortion in one direction with the reciprocal of the scale distortion in another direction, the areas in many locations on an equal-area world map tend to appear squashed and have unsatisfactory outlines (as Japan would in Figure 2.1).

Although projections are mathematical in nature, the visualization of some projections, as exemplified by Figure 2.2, is common and can help in understanding how they are formed and what some of their properties may be. For example, it is clear that for the azimuthal projection shown (and indeed for all azimuthal projections centred and tangent at a point) the projection onto the map of bearings on the globe from the tangent point to any other point maintains the true value. However, just because some bearings are correctly represented, the map is not necessarily conformal.

Any projection can take only one of the above three forms (conformal,

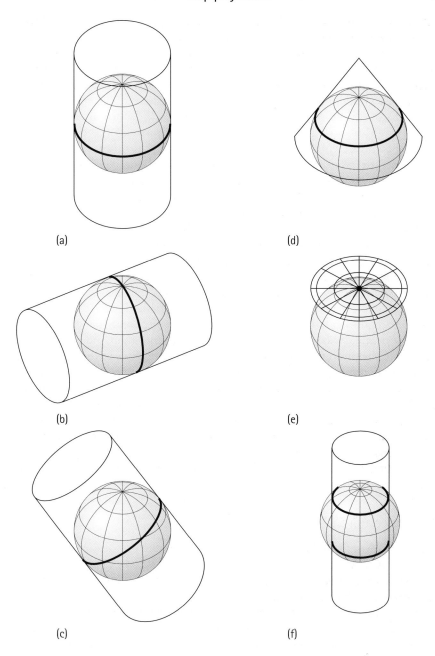

(a)

(b)

(c)

(d)

(e)

(f)

Figure 2.2   Geometric visualization of projections: (a) cylindrical (normal aspect); (b) transverse cylindrical; (c) oblique aspect cylindrical; (d) conical; (e) azimuthal (with bearings on earth projected to bearings on the plane); (f) 'secant' cylindrical (normal aspect) (tangent lines and points – lines of zero distortion – are emphasized).

equivalent or equi-distant), but there are further parameters associated with projections which can be varied to customize the appearance of a map. These include the aspect, the centring, and the area of coverage. With reference to Figure 2.2 it can be noted that the plane, cone or circle used as the surface onto which the graticule is 'projected' can be positioned in a number of different ways. In the 'normal' aspect of the cylindrical projection the cylinder is positioned tangential to the equator (the tangential line is the line of zero scale distortion). A transverse aspect cylindrical projection results when the main axis of the cylinder runs east–west and the cylinder is tangential to a meridian. Any other arrangement of the cylinder (or indeed the cone or plane) is termed oblique. The centring of the projection can be varied to suit the purposes of the map-maker (see Figure 8.5), with, for example, an azimuthal projection being made tangential at a certain point (and thus having zero scale distortion at that point and true bearings from that point). For instance, an airline map portraying the true directions (bearings) from Schiphol Airport, Amsterdam, would be centred at 52° N, 5° E.

The mathematical basis of projections ensures that they are amenable to computer assistance in their preparation, and in some cases the mathematical transformations are trigonometrically straightforward (although in just as many cases they are extremely complex). Computer-assisted methods can overcome the tedious manual calculation and plotting of map projections, the graticule and mappable detail; although it is important to realize that the rapid drafting and changing of the projection of a map has only become possible in the last 20 years of a science that measures its history in millennia. The assumption (so far maintained) that the earth is a true sphere further eases the calculations. The transformations that are embodied in the mathematical functions employed in map projections can be manipulated in an infinity of ways, resulting in a host of arrangements of the graticule, including 'interrupted' projections (Figure 2.3). Knowledge of the transformations and their parameters means that it should always be possible to 'reverse-engineer' the projection process and compute true geographical coordinates (latitude and longitude) from projection values.

Map projection interpretation and choice

There are thus various different properties which a map projection may possess and which it may be necessary to apply when a map is to be created. Similarly, it may be thought possible to ascertain the type of projection used on an existing map by referring to the graphical portrayal of the graticule and the nature of the lines that form it. Considerations of whether the meridians or parallels (or both) are curved lines or straight, whether the spacing between the meridians and parallels is constant or increases towards the equator or poles, and what the shape is of the outline of the map (rectangular, circular, etc.) can give hints to the projection used. Unfortunately there are so many further variables, not least the possible arbitrary definition of the mapped area by imposed neat lines (sheet lines which bound the mapped detail and define the edge of the map), that it is often impossible to determine the true type of projection used without a statement by the cartographer on the map face.

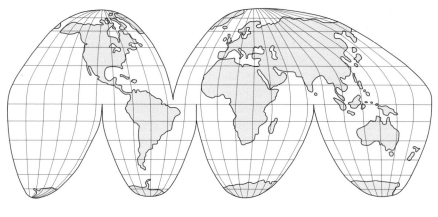

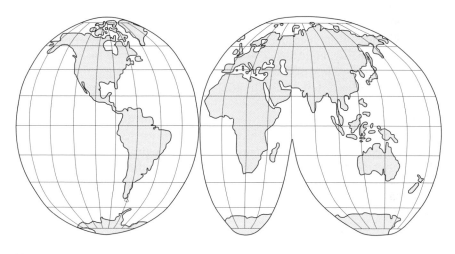

(b)

Figure 2.3 Examples of 'interrupted' map projections of the globe: (a) Goode's; (b) interrupted Molleweide.

It is often a requirement to know the type of projection presented when the map is being used as a working document, e.g. in navigation, or when data from the map are being combined with other spatial information from a different source. The compilation of computer databases of spatial locations over a wide area from a variety of different source maps, which may be on differing projections, is common in geographical information systems (GIS; see Chapter 7). The projections of these diverse data sources need to be known and such systems therefore often have extensive facilities to compile, transform and represent spatial data in a wide variety of map projections.

The choice of which projection to cast the graticule and geographical outline onto is governed by a host of factors relevant to the map-making task in hand.

The scale of the mapping, as will be indicated on p. 38, can determine whether or not the area of the globe to be portrayed is small enough for it to be treated as a plane, flat surface. If a large zone is to be mapped, its location on the earth's surface may be of importance. Polar areas are traditionally projected using azimuthal projections, whilst mid-latitude lands with significant east–west extent may benefit from conical projections that have lines of zero distortion along one or two parallels. Zones with a large north–south dimension (e.g. countries such as Chile and Sweden) are ideally presented using a transverse cylindrical projection. In effect what is required is minimal distortion on the map. If distortion can be pushed to the edges, if it can be re-distributed using an interrupted projection, if it can be evened-out to a certain extent by using more than one line or point of zero distortion (as on a 'secant' projection; see Figure 2.2), then the map-maker will usually decide to make those choices. This is because distortion is seen as an anathema to 'good' cartography. Cartographers may differ in their view of which distortions are worst; although we have shown that it is an inevitable part of the map-making process, even of forms of alternative cartography described in Chapter 8.

The purpose of the map may determine which of the fundamental properties of conformality, equivalence and equidistance are to be maintained. For some applications, none of these is necessary and there are a number of projections that have been designed with aesthetic appearance, rather than conformance to a mathematical property, as their major merit. Such projections include the Times and Robinson projections, used in many world atlases; a series of perspective projections which represent the earth's surface as if viewed from a spacecraft; and various 'magnifying glass' type projections which drastically enlarge the centre point of interest compared to the margins of the map. Further radical reorganization of the layout of the globe is evident in the construction of 'cartograms', and other 'alternative' maps further described in Chapter 8.

Distortion

A full treatment of map projections and their properties relies on a familiarity with mathematical calculus. Fortunately, however, the application of *Tissot's indicatrix* (an ellipse of distortion which can be mathematically constructed for every point on the earth's surface) is graphically straightforward and this can give a ready insight into the properties of a map projection, as indicated in Figure 2.4. The construction of the ellipse at a point, of which a circle is a special case, relies first on a knowledge of the particular, possibly varying, scales along the meridian and along the parallel, and secondly on the orientation of the *orthogonal principal directions* (those directions which are perpendicular to each other on both the globe *and* the projected plane at that point). The former are represented by the magnitude of the major and minor axes of the ellipse, whilst the orientation of the ellipse denotes the maximum angular deformation at that point. If the ellipse is circular throughout, the projection is conformal; although it will vary in size, indicating that a conformal projection can

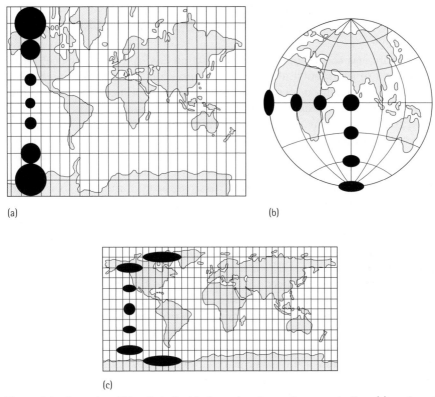

(a)   (b)

(c)

Figure 2.4 Examples of Tissot's indicatrix for various types of map projection: (a) conformal (cylindrical Mercator projection); (b) equal-area (transverse azimuthal); (c) equidistant (cylindrical Plate Carrée).

never be equal-area. If the ellipses have the same area throughout, the projection is equal-area. In this case, the ellipses will not have a common shape and they will be distorted with differing major and minor axes. An equidistant projection has the minor axis of each ellipse equal in length to the diameter of the circle at the line or point of zero distortion (Figure 2.4). The ellipses are determined from the mathematical functions which define the projection and it should be noted that, theoretically, they are infinitesimally small (Figure 2.4 exaggerates their size immensely to show their localized appearance): each point on the surface has a different ellipse calculable at it. Maling (1993) gives a worked example of the application and interpretation of Tissot's indicatrix and its parameters for the Hammer–Aitoff projection.

Larger-scale mapping

At the very large scales of map-making undertaken by engineers and surveyors, it is often assumed that the portion of the earth being mapped is flat. Such an assumption can be made because the distortions induced by mathematical projection are minimal over small areas. Thus, measurements made in the field can

Box 2.2   Personality box – The 'Peters projection' and its instigator

The debate over the use of the Peters projection has been a notable example of the impact of a particular ideology on the subject of the mapping, and hence the map-making, process. The recognition of the role of societal and individual *mores* in practical cartography had been expressed as early as the nineteenth century by some German cartographers: it was further developed in the Anglo-American cartographic tradition by geographers such as John K. Wright. However, it was the controversy over the use of the Peters projection that initiated the most widespread examination of cartography as a social, and indeed political, rather than necessarily a scientific (and therefore, it is widely assumed, objective and ideology-free) activity.

Arno Peters (1916–) certainly followed a political agenda in developing and promoting 'his' projection. As a historian he was aware of the imbalance in the world-view in most accounts of the development of the human race. In particular, historians took a Eurocentric bias into their writings, and this attitude was both promoted and reinforced by the graphical world-view presented by the commonly used Mercator projection (see figure).

Peters' intention in developing a new projection was to create an equal-area projection which overcame some of his objections to Mercator's (despite the fact that a large number of equal-area projections were already known). Although there was a claim that the development of this projection took ten years, it is, in fact, an extremely simple and straightforward transformation (the mathematical functions used are $x = $ (scale / $\sqrt{2}$) × longitude; $y = $ scale × sine (latitude); see p. 28 for an example of how such mathematical functions are applied). The result, however, was regarded as novel enough (see figure) and Peters' propagandizing on its behalf was persuasive enough, for the world map, as represented on 'his' projection to be adopted as an official icon by organizations such as UNICEF, Christian Aid and the 1980 Brandt Report on international development. In addition, Peters created a global atlas using the projection with world coverage consisting of 43 equal-sized maps (each portraying 8 750 000 km^2) plus a section of thematic maps of the whole world.

In fact, Peters' world map is still highly conventional, usually centred on a meridian close to Greenwich, rectangular in outline, with north at the top and with the poles unrepresented (although two azimuthal projections – using Lambert's equal-area projection – were later added for the atlas). His desire to be fair and balanced to the peoples of the world is at odds with an equal-area projection which gives as many atlas pages to the sparsely populated Siberian tundra as it does to the whole of China and India combined. The true value of Peters' projection may, therefore, be opposed to his claims to represent the peoples of the world fairly. The fact that four pages of this atlas are needed to cover the Sahara, whilst only two pages are necessary to contain most of the countries in Europe is evidence of a dramatic shift in emphasis

(continued)

(continued)

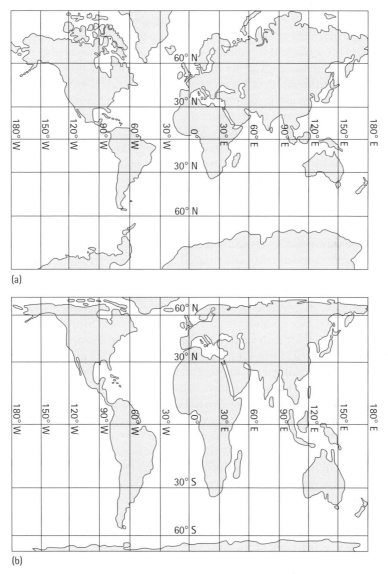

(a) Mercator's projection of the globe; (b) 'Peters' projection' of the globe.

from traditional atlases which, although concentrating on powerful states at the expense of the Third World, at least scaled down the representation of unpopulated areas.

Such shortcomings, along with more detailed technical irregularities and false claims made on behalf of the projection and its originality (it is, in fact, a direct replica of a projection devised in 1871 by Scottish clergyman James Gall),

(continued)

(continued)
led to a strident campaign against the use of his projection, and against Peters himself. This assault was led by many cartographic authorities, particularly in academic circles. As with many of the personalities highlighted in this book, Peters was an excellent self-publicist who claimed that his view of the world had merit beyond that which some others suggested it was reasonable to claim. What the controversy over his projection did do, however, was to re-awaken interest in world mapping, improving awareness of the nature of people's inherent mapping 'urges' and their impact on the art and science of map-making. In particular, it opened up cartographic representation to social and political scrutiny in a way that had not been done before. Although many cartographers still prefer to focus solely on the graphical and mathematical shortcomings of Peters' work, others have been receptive enough to take on board the message that there is merit in considering different world-views.

be scaled and transferred directly to a Cartesian (x, y) coordinate system (or grid), with little likelihood of errors in fitting all the components and features together.

Often, however, large-scale maps form a uniform series covering a large area (often these are topographic maps, as described on pp. 90–93) and in this case, the production of accurate maps based on precise measurements taken in the field over a large area involves the cartographer in considering the true nature of the earth's shape. Not truly spherical, the earth is an irregular solid with considerable variation in its form (in effect, it is flattened at the poles, with perturbations over the remainder of the surface). The mathematical transformations (projections) that are used for such map-making rely not on a sphere as the generating solid, but on a mathematical approximation (known as the *spheroid* or *ellipsoid*) to the irregular, slightly pear-shaped shape (the *geoid*, equivalent to mean sea-level over the earth's surface) which is the true figure of the earth. The approximation used will vary with location on the earth's surface as it is a mathematically defined portion of the 'best fit' spheroid to the geoid at that point. This curved surface can then be transformed to the plane surface using similar (although always more complex) mathematical functions to the projections described above. The parameters (i.e. the centre and major and minor axes) of such ellipsoids are optimized for local conditions and thus different ellipsoids are used in varying parts of the world.

Moving to the extremely large scales sometimes used to create maps of urban areas, for example, or of individual buildings, the world can again not be assumed to be flat and mapping clearly becomes an exercise in representing the three spatial dimensions that we inhabit. Such representations are considered further on p. 111.

Generalization

A major reason for representing the world at differing scales is to show varying levels of detail. Clearly a map at a scale of 1:100 000 only has one-quarter the

 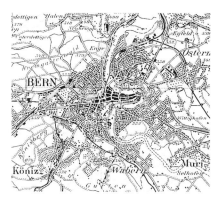

Figure 2.5A Examples of generalization: generalization due to scale reduction (Bern, Switzerland at 1:25 000 and at 1:100 000 scale).

space available for portraying a particular area compared to a map at 1:50 000. From a traditional paper-based cartographic perspective this space restriction is a fundamental determinant of the content and appearance of the map. Computer-assisted mapping on a monitor screen, which can exhibit 'zoom' facilities, does not suffer from the same rigidity, but there is still a limit to what can efficiently, aesthetically and definitively be represented at any particular scale. The object of cartographic generalization is the production of such an efficient, aesthetic and definitive image. Every map is a generalization of reality and every map must take generalization into account to reduce complexity, to retain accuracy and to counteract some of the undesirable consequences of scale reduction.

Unlike the mathematical rigour of map projection, the generalization which transforms the impressions and properties of the real world to a map is extremely and explicitly subjective. Decisions on what features to include, whether to exaggerate, displace and classify them, and how to symbolize and simplify them, are not governed by strict rules, although considerations such as the intended audience for the map, the scale of representation, its purpose, the nature of the area and theme being mapped, and any design restrictions, can help to determine the generalization (Figure 2.5). It is at this stage in the map-making task that the cartographer has most freedom, and it is essential that the map user be aware of the fact that the map portrays one (or at most merely a few) person's view of the world. That person will have their own prejudices and ignorance, will be subject to political and social influence which may create a 'hidden agenda' and govern the representation, will have a historical background which brings certain ideas and norms to the fore of the mapping process, and will be working within a particular organizational structure which may affect the map content and design.

For these reasons, it has been impossible to give anything other than general guidelines about generalization to the map-maker. Such advice is often to observe the map purpose, to maintain the essential character of the earth's

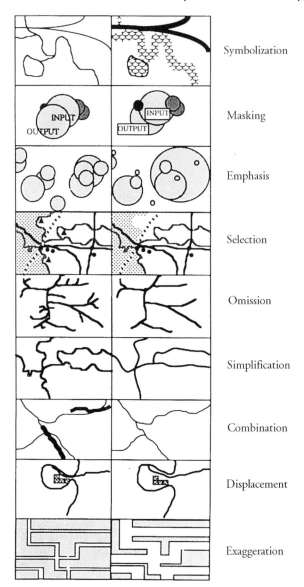

Symbolization

Masking

Emphasis

Selection

Omission

Simplification

Combination

Displacement

Exaggeration

Figure 2.5B Examples of generalisation: the variety of generalisation operators (from Mackaness, Chapter 13 in Buttenfield and McMaster, 1991).

surface in the area being mapped, and to be consistent over the whole of the map. Despite its intangible nature, however, cartographers do need to be able to recognize when to undertake generalization, and how to undertake it. In recent years a great deal of research effort has been spent trying to automate certain aspects of cartographic generalization by creating computer-compatible algorithms. Progress has been slow, however, perhaps because the inherent nature of the process is not mechanical.

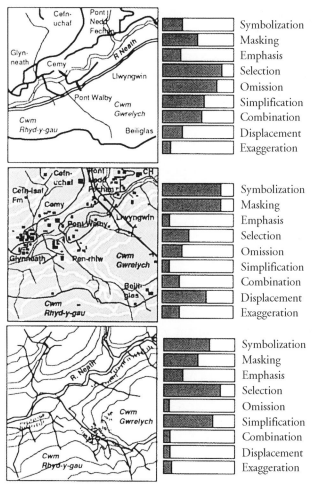

Figure 2.5C Examples of generalisation: using operators for purpose-varying generalisation – car touring, walking and terrain illustration (from Mackaness, Chapter 13 in Buttenfield and McMaster, 1991).

Recognizing and resolving the need for generalization

Generalization becomes necessary when the features that we wish to portray on the map start, due to the data compilation and representation process or due to scale reduction of existing data, to coalesce, overlap and conflict with each other. Such a situation can be determined by eye when creating a map manually: it can be recognized more objectively by calculations of such parameters as the distance between features (which can be tested against desirable thresholds of separation), the density of point features, the sinuosity of lines and the compactness of areas (measurable by indices such as the ratio of area to perimeter). The resolution of such conflicts and the clarification of the graphical image involves operations such as *simplification* of the features, *selection* or *omission* of detail, *exaggeration* of size (which may result in *displacement* of other features), *combination* of similar features, replacement of the graphical representation of an object by a changed or simpler *symbol, emphasis* of some objects and possible *reclassification*, along with the potential *masking* of detail to improve clarity (Figure 2.5). Although this list is a discrete set of logical operations, generalization affects all symbols on the map simultaneously and should therefore be approached as an holistic exercise. The breakdown of generalization into these steps, however, can ease the creation of the computer-coded algorithms mentioned above, which can attempt generalization in a computer-assisted mapping system and can possibly increase the objectivity and consistency of the inherently subjective process. Such algorithms do not, of course, make the process completely objective (they have to be applied under human guidance in an interactive digital mapping system or abide by human-initiated rules within a computer-based 'expert system' which stores 'artificial intelligence') but they do make the operations more explicit and hence more consistent and 'scientific'.

Summary

The questions of choice of map projection and of map scale and the consequent level of generalization of detail are important at the initiation of any map-making project, and they are preliminary indicators that map-making is a subjective process, with considerable choices available to the prospective map author. Variations in extent of coverage, in content, in design and appearance, in emphasis of features on the map face, in the medium on which the map is presented and in method of production, are all capable of reflecting the world-view of the map-maker, but also of inducing differing world-views in the map user. The immediate impact of the scale and projection and the effects of generalization are among the most important of these variables.

Further reading

A well-illustrated description and interpretation of the pictorial and diagrammatic maps used in advertising and ephemeral material is found in *Pictorial Maps*, by N. Holmes (Herbert Press, London, 1992). A survey of the most widely used map projections for small-scale maps is given in F. Canters and H. Decleir, *The World in Perspective* (John Wiley, Chichester, 1989) which also has an accessible description of the distortion inherent in all projections. *Coordinate Systems and Map Projections*, 2nd edition, by D. Maling (Pergamon, Oxford, 1993) is a complete and respected manual on all aspects of map projections, including their calculation and plotting. It also presents the Tissot indicatrix diagram for the sinusoidal projection (p.115). The history and development of map projections is covered in authoritative detail by J. Snyder in *Flattening the Earth* (University of Chicago Press, Chicago, 1993). Arguments against the use of rectangular map projections are given in 'The Case Against Rectangular World Maps' (Anon.), *The Cartographic Journal*, **26** (2)(1989), 156–157. The

Peters controversy is dispassionately discussed in Chapter 1 of M. Monmonier, *Drawing the Line* (Henry Holt, New York, 1995). The centrality of generalization in the mapping process, along with an explanation of its practical implementation in both manual and computer-assisted cartography is given in the collection entitled *Map Generalization: Making Rules for Knowledge Representation*, edited by B. Buttenfield and R. McMaster (Longman, Harlow, 1991).

Chapter 3

Navigation, maps and accuracy

Introduction

Although none of the maps prepared by Ptolemy have survived (see Box 1.2), it has been possible to re-create his view of the shape of the world by referring to the gazetteers which form the major part of his work, *Geographia*. The locations of towns and geographical features presented in such gazetteers ensure that maps displaying the positions of these features can be constructed. All maps rely on such locational information as their raw material: absolute locations of topographical features (e.g. their latitude and longitude) and relative locations of neighbouring locations (e.g. the lines and sequences of stations on a map of an underground rail system). Some cartographers would suggest that it is only those graphical images which show, as far as possible within the limitations of scale and projection, the 'true absolute position' that can be called maps. Products such as the London Undergound map, which is a 'linear cartogram', showing relative positions only of stations along branch lines, should perhaps be termed diagrams (Figure 3.1).

Although they may not show true coordinated position, diagrams such as linear cartograms do have a major role to play in communicating information about the spatial locations of features. In particular, they are often designed to show *topological* relationships. These are connections which, although not necessarily represented *geometrically* correctly (i.e. it may not be possible to measure the scaled distance from the diagram), reflect the proximity and contiguity of locations. Because a linear cartogram maintains linear topology, it can be determined from Figure 3.1 that St John's Wood is one station away from Baker Street on the Jubilee Line, but it is not as close as the Underground map implies. In many map-use tasks, such as navigating along a linear route such as a freeway or suburban rail line, the topological relationship is all that is required and *strip maps*, a particular type of linear cartogram (see Figure 3.2 for an example), can be used.

Positioning

The majority of navigational tasks for which maps are required are much less restricted (both in the possible range and in the possible direction of travel) than merely determining which freeway exit to take. A motorist may want to be able to consider alternative cross-country routes; the military pilot is likely to fly well away from the air traffic corridors along which his civilian counterpart moves; and the recreational mariner wants an overview of the entire bay in which she is sailing. In these circumstances, of course, a traditional map product is required in the form of a road atlas, a tactical pilotage chart or a hydrographic chart, to assist in the navigation. These representations of the world require a high degree of precision in the absolute positioning of features in order to avoid mistakes in map use such as taking a wrong turning on the road, flying into a mountain or grounding on a reef.

Methods for determining an accurate position on the surface of the earth are therefore a fundamental prerequisite for all mapping. The history of cartographic representation has been inextricably linked with the technology of determining position on the earth's surface. Thus techniques of field astronomy, land surveying, hydrographic (sea) surveying, mapping from aerial photographs (photogrammetry) and from satellite imagery (remote sensing) have all been useful in the creation of map frameworks and detail which reflect accurate positions.

The current preferred method of position determination, both by land surveyors collecting spatial data and by navigators and many field scientists who wish to locate themselves accurately, relies on a system of artificial satellites orbiting the earth (as distinct from the stars by which ancient peoples often navigated). As with the stars, the positions of the satellites are known or predictable, but instead of being visible (which is necessary for navigation by stars) the satellites broadcast their presence and position by sending out radio signals. One constellation of such satellites forms the Global Positioning System (GPS) which is now relied upon by surveyors and travellers world-wide to determine their location (see Box 3.1).

Navigation

Reliance on stars for position fixing was the foundation for successful navigation by the earliest seafarers in the Mediterranean. The tradition of the Phoenicians was continued by successive generations of navigators in the region. They were aided by developments in cartography such as the introduction of portolan charts (see p. 17) which themselves relied on data from astronomical observation, on dead reckoning (positioning oneself by using time and average speed to estimate distance travelled, and noting the direction of travel) and on elementary shipboard compass survey for their creation.

Charts are used by navigators in conjunction with route planning methods such as compass bearings. It is helpful if such bearings (or azimuths), which are plotted on charts and followed by mariners, are rendered as straight lines on the charts. In order for this to happen, the chart must have a conformal projection (see Chapter 2). Note that a journey on a constant bearing is not the shortest distance between two points on the surface of the earth (unless one is travelling due north or south or along the equator). A route along a great circle, which does define the shortest physical distance between two points, has a constantly changing azimuth (except for the cases just mentioned) (Figure 3.3).

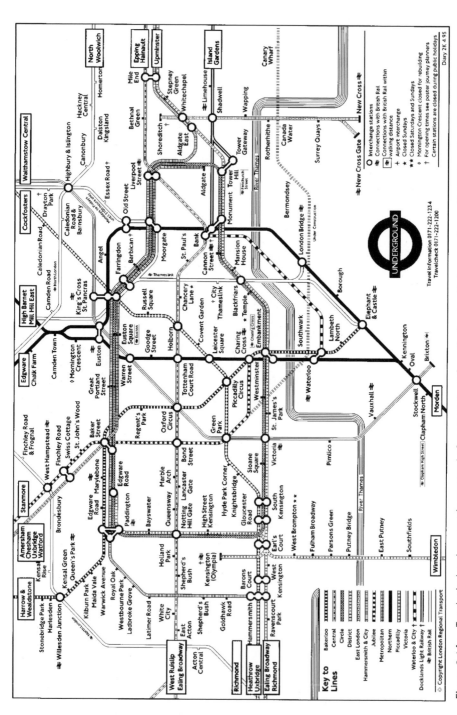

Figure 3.1A: Topological and geometric portrayal of space: extract from the London Underground map. © LRT Reg. User No. 97/2482.

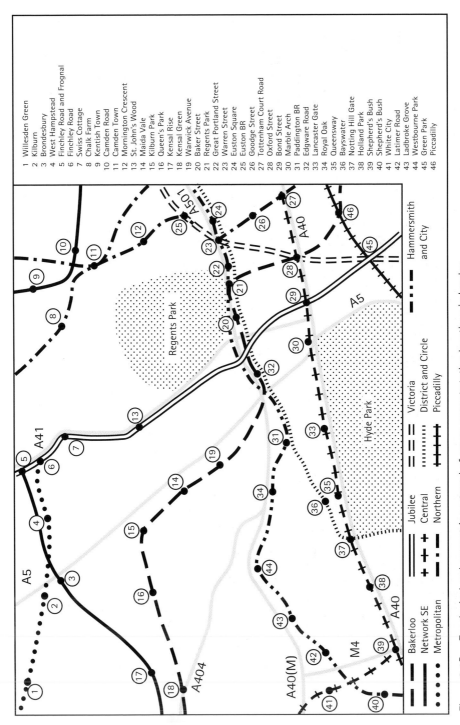

Figure 3.1B: Topological and geometric portrayal of space: correct station locations in London.

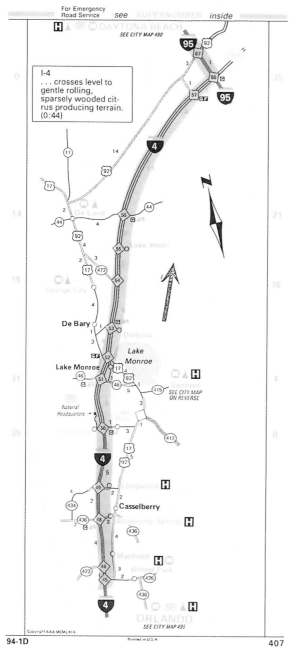

Figure 3.2 An example of a strip map: an American Automobile Association Triptyk, showing a road route. © AAA reproduced by permission.

Box 3.1   Contemporary mapping box – GPS and positioning

The revolutionary abilities of GPS to determine position (latitude, longitude and height) to high standards of accuracy, day or night, in all weather conditions, and the remarkable openness of the US government (which has developed and owns the system) to its use by civilians throughout the world has led to its application in a variety of fields which involve geographic data collection and presentation. GPS data collection can be used for:

- logging attribute and positional data in the field;
- supplementing technology in areas such as in-car navigation and fleet control;
- revising and maintaining computer-based spatial data sets;
- interacting with cartography by
  — updating position on dynamic map displays;
  — allowing for speedy and accurate creation of spatial data sets for digital mapping.

The GPS constellation consists of 21 satellites (plus three spares) which orbit the earth on precisely calculable paths, broadcasting highly accurate time and locational information. Standard GPS receivers, which range from hand-held devices to considerably more precise instruments costing tens of thousands of pounds, need to be able to sample signals from at least three satellites. There is a range of techniques which can be applied by the user to improve the precision of GPS positioning. Some of these try to overcome the potential and actual errors that affect GPS – unintentional errors in the satellite itself (faulty clock, incorrect orbit broadcast), intentional degradation of the signal by the US military authorities, and delays in the path of the signal through the atmosphere from the satellite to the receiver. Other data manipulations are undertaken to sample different components of the signal, to use signals broadcast from known or intermediate points on the earth's surface to refine the raw position obtained, and to operate GPS in real time. In all cases the intention is to improve the accuracy of the position obtained. Estimates of positional accuracy vary, but civilian users accessing the coarse acquisition (C/A) code (the usual signal sampled by civilian single-frequency receivers) can expect to get within 100 m of their true planimetric position 95% of the time. High-quality repeatable positioning at the centimetre level of precision is possible with the most sophisticated equipment and techniques.

The use of GPS is evident for all types of map data suppliers and users: foresters can undertake 'field digitizing' to obtain boundaries of remote stands of timber without difficult land survey or photogrammetric techniques; field scientists can quickly obtain an accurate position of the location of samples without needing to interpolate a map; asset managers can record the attributes of their properties or facilities and build up extensive urban data sets merely by 'walking the streets' and recording positions.

(continued)

(continued)

At the same time as it can make many aspects of work, and also leisure and warfare (considered in Chapter 5), more efficient, concerns have been raised about the exploitation of resources and people using this new technology. Such concerns can be related to the economically efficient, but ecologically unsound, harvesting of rural resources, the exploration for minerals in the most inhospitable places and the 'tagging' of human beings to know their location at any time. All of these activities can be undertaken using GPS.

From a technological perspective, however, the world-view of the GPS user is closely related to the physical nature of the earth. Most large-scale and topographic maps assume the earth's shape to be spheroidal and although GPS latitude and longitude readings are read from receivers based on the WGS (World Geodetic System)-84 spheroid, they do need manipulation to obtain grid co-ordinates on other spheroids and projections for mapping purposes. In addition, the interpretation of height information from raw GPS data can be problematic, as the geoid, to which such information is referred and which is in effect the shape of the mean sea-level surface around the whole world, does not coincide with the spheroidal surface. Although positioning and navigation can be undertaken with care using GPS in real-time, map production and GIS data set generation need to be approached with a knowledge of the limitations, as well as the opportunities, of GPS. GPS receivers are too often seen as accurate 'black boxes' always giving the 'right answer': an appreciation of the methods used in satellite geodetic surveying is necessary for any map-making task using positions derived from GPS.

The most prevalent conformal projection is that attributed to Mercator, and it is the Mercator projection that has been used on the majority of sea charts. Air charts often use other conformal projections, e.g. Lambert's conformal conical. In neither case was the projection devised for representing the entire earth, and the use of Mercator, in particular, to show the whole world has long been the subject of criticism (see Chapter 2). Mercator, despite being a productive cartographer, never produced a world chart using his own projection (Box 3.2).

The navigator's world-view: chart content, travel speed and the skills of navigation

The term 'chart' is used for many of the graphical representations used for navigation (although there are road *atlases*, orienteering *maps* and sea *books*). Today, most of these products are in a transitional stage whereby the paper copy is being replaced by digital equivalents. This is especially true for seaborne navigation (see Box 3.3), but progress is also rapid inside many road vehicles (see Box 3.4) and in the aircraft cockpit (see Box 3.5).

The navigator journeys across the surface of the earth using either direct viewing of the surroundings or with reference to navigational signals. The chart

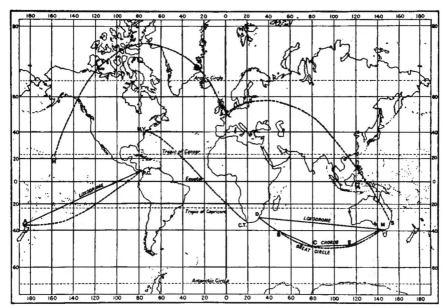

Figure 3.3 Lines of equal bearing (loxodromes) and Great Circles on the earth's surface, on Mercator's projection.

therefore needs to portray the features of the landscape or seascape and/or the nature and location of navigational aids such as buoys, radio beacons, road junctions and airport control towers. The content of charts, and the view of the world they represent, is governed by the immediate and continuous need for specific spatial information, easily accessible and understandable whilst on the move. For this reason, charts tend to be sparse in terms of detail (Figure 3.4). In addition, they are working documents, often annotated by the navigator who needs space to plot bearings and mark waypoints (places where the route is changed to follow a different bearing). Furthermore, a unique feature of sea-charts in particular is that the revisions and updating of the graphical image are traditionally done by the users themselves. A sailor is used to receiving regular 'Notices to Mariners' in text form, describing new features (e.g. ship-wrecks) and changed objects (e.g. altered lighthouse illumination sequences). Such changes must be incorporated graphically, and the fewer items there are on the chart, the less onerous and less error-prone this operation becomes.

Land-based navigation involves similar considerations of map creation and map use. Modern-day road atlases have specific design requirements: normally the road pattern and numbering are the boldest items, often in bright colours such as red or orange, allowing for easier perception at speed. Many other features that may be visible to the motorist during the journey are often omitted, e.g. the terrain, the hydrology and river system, the railway lines and the surface vegetation. In effect, the content is governed by the speed at which the motorist travels across the landscape and the perceived level of important detail which it is expected that the road user will need. Once the speed of travel decreases (e.g.

Box 3.2   Personality box – Mercator and his world–view

Mercator (1512–1594) was born Gerhard Kremer near Antwerp, although he did most of his work as a refugee working for a ducal patron in Duisburg in present-day Germany from 1552. He started work as a land surveyor, undertaking all the measurement as well as the map production stages for an accurate map of Flanders published in 1540. He then became a mathematical and astronomical instrument-maker before becoming intrigued by cartographic representations of larger areas. His map of Europe on six sheets was finished in 1554. The influence of Ptolemy was continuing to be felt by map-makers of this time and Mercator published his own re-drawing of the maps from *Geographia* and the text of the work itself. He used some of its material in his map-making and, although over successive globe and map products he gradually reduced the size of the Mediterranean Sea from its Ptolemaic excess, he still made its east–west extent too large. Much detail on Mercator maps of areas outside Europe was also very distorted despite being long explored (e.g. the Black Sea was too large).

Mercator gradually extended the range of his map-making, although he concentrated on relatively large-scale maps of continental Europe and found time to help other cartographers such as Ortelius and William Camden establish their reputations. He produced his maps on a variety of standard projections, but he had an innovative influence on map design, popularizing italic style lettering in his engravings, for example. The first collection of his maps to be published in book form covered France, Germany and the Netherlands in 51 separate map sheets. Further volumes followed in 1590 (Italy and south-east Europe) and posthumously in 1595 (northern Europe and the rest of the known world). This last volume was the first collection to be titled an 'atlas'. Mercator's maps were extremely well received and the copper plates on which they were engraved were in use, having been sold, revised, doctored, plagiarized and falsely published as new products (all of which were common practices during the flowering of Dutch cartography in the seventeenth century) for many decades after.

Mercator's achievement in developing the map projection that bears his name has overshadowed his many other accomplishments. Published in 1569, the projection is conformal and was specifically designed to aid seaborne navigation. It is not known exactly how Mercator created his projection, but it is likely to have been by graphical means, either by transferring bearings from a globe to a flat sheet or by plotting the positions of the intersections between meridians and parallels using simple geometric constructions. It was left to an English mathematician, Edward Wright (1561–1615), to develop the simple trigonometric relationships allowing for the creation of precise plotting tables of projection coordinates for each latitude (Snyder, 1993).

The Mercator projection rapidly became the accepted representation of the oceans for navigators: Captain James Cook used Mercator-projected charts for his extensive explorations around Australia and in the Pacific in the eighteenth century, and the projection is still used on most hydrographic charts of large

(continued)

(continued)

and medium scale. Scientific mapping of the oceans has also used Mercator: Edmond Halley produced a map on this projection of magnetic declination in the Atlantic Ocean in 1701; Benjamin Franklin contributed to a chart of the Gulf Stream in the North Atlantic Ocean in 1770. Unfortunately, the use of Mercator-based mapping for the land areas of the world (where navigation does not rely on magnetic bearing to the same extent) is not suitable but, as indicated in Chapter 2, it became widely used especially for Eurocentric world mapping in the nineteenth and early twentieth centuries. That trend has since been reversed, amidst inappropriate and strident criticism of the Mercator projection, which is in fact eminently suitable for the task for which it was developed – seaborne navigation in the mid-latitudes.

during a search for a specific suburban address), the level of detail of the map used for such a task increases: a street directory has the name of every road, along with information related to important local landmarks, buildings, parks, shops and pubs.

The speed of the airline pilot is obviously many times that of the private motorist and air-charts can similarly match swiftness of travel with sparseness of detail. For civilian pilots, few of whom fly by 'looking out the window', the configuration of the landscape is of lesser concern than the location and nature of electronic navigational aids. Their need for paper charts nowadays is, in fact, fairly limited: those that are produced emphasize air traffic corridors, locations of beacons and radio frequencies and call signs. The military pilot, on the other hand, who may be flying at low altitude and relying on vision to a large extent, may require charts with more topographic detail, especially relating to safe flying heights in the vicinity and potential obstructions such as power lines and tall chimneys. An example extract from such a chart is shown in Box 5.2.

Navigation is an integral component in asserting human beings' dominion over the surface of the earth, and the maps and charts that have been produced to enable navigation can be seen as devices to exert control over space. The recent impact of technological change in navigation is far-reaching and, to a certain extent, has diminished the role of graphical presentations in the navigational process. Today it is possible to key destination coordinates into a GPS-based navigational tool and let the instrument determine current location and optimum direction and route to that position. All the navigator need do is to steer her craft in the direction indicated. Traditional skills in position determination, route planning, compass setting, direction following and map reading are all redundant in this scenario.

Accuracy

For reasons of safety (and to meet certain international treaty obligations, such as the Safety Of Life At Sea (SOLAS) Conventions 1974), it is clear that the information portrayed by chart cartographers needs to be of the highest

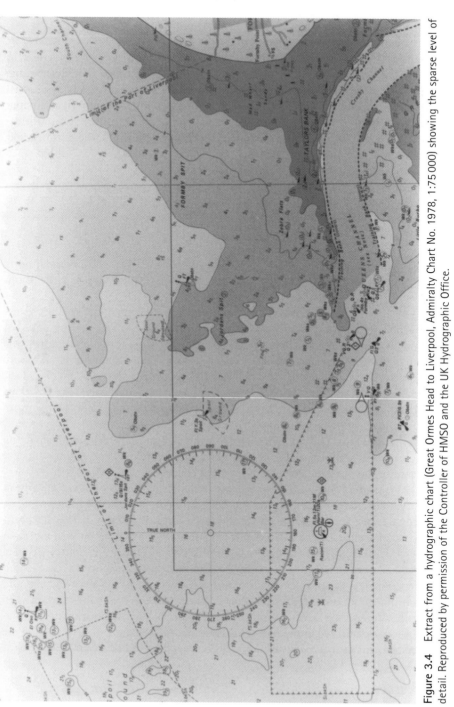

Figure 3.4 Extract from a hydrographic chart (Great Ormes Head to Liverpool, Admiralty Chart No. 1978, 1:75 000) showing the sparse level of detail. Reproduced by permission of the Controller of HMSO and the UK Hydrographic Office.

Box 3.3   Contemporary mapping box – ECDIS

The Electronic Chart Display Information System (see figure) is the world-wide implementation of agreed standards, put forward by the International Maritime Organisation, for supplying digital positional and navigational information to the captain of a ship. It is thus intended to replace the paper chart, carried (by maritime law) by all sea-going craft for centuries. The advantages of displaying spatial information digitally on the ship's bridge are similar to those put forward for digital mapping in general (see Chapter 7):

View of ECDIS in use onboard a ship. Reproduced by permission of the Controller of HMSO and the UK Hydrographic Office. © Crown copyright.

- selectable detail, e.g.
  — show the shipping channels and navigational aids, but not the soundings or the coastline.
- changeable design, e.g.
  — render the detail in colours appropriate for night vision, as opposed to the traditional design scheme which is more suitable for daylight chart use.
- editable updates, e.g.
  — allow for the incorporation of a new lightship symbol at this location.
- ease of use, e.g.
  — display an enlargement of this area on the monitor;
  — portray at large scale a feature that was once at the junction of two small-scale charts;
  — give a generalized overview of the area for which the paper chart is an unwieldy size;
  — pan across to a further area which was, in the past, on a totally different paper chart;

(continued)

(continued)
   — allow the superimposition of a small-scale route diagram onto this large-scale display.
 • navigational integrity, e.g.
   — sample the on-board GPS receiver (see Box 3.1) and constantly plot the changing position of the ship on the display;
   — predict and display a route and position given the current speed and direction and the prevailing sea conditions;
   — give an audible warning if the ship breaches or approaches a known hazardous feature such as a 10 m depth curve or a buoy.

ECDIS is therefore more than the digital equivalent of a paper chart; it incorporates a digital hydrographic database of oceanographic and navigational information, a system for incorporating updates, a display device for computer-assisted map presentation, a notepad on which local knowledge, prevailing conditions and route planning can be handled, and a vital part of the navigational and positioning system on board.

The legal status of the ECDIS system is still unclear, and it is debatable whether a ship that carried no paper charts at all would be covered by maritime insurance. However, considerable developmental work on electronic charting is being undertaken by a large number of national hydrographic mapping organizations, and very soon the majority of large sea-going craft will use digital charts regularly.

accuracy, both in terms of the position and the nature of features. This is also the goal of many other cartographers, whatever map they may be working on (although there is often deliberate distortion of maps (for example, see p. 147) and subconscious distortion, which is endemic in any creative activity, including cartography, and is considered throughout Chapter 4). Cartographers therefore spend considerable amounts of time assessing the accuracy of the source material which they collect for subsequent map production. A further aspect which is considered at this compilation phase is the relative accuracy of various map components: it is important to ensure that the accurate display of one set of features does not lead to unacceptable distortions in others. The traditional nature of cartographic production is such that there are continuing attempts throughout the map-making processes to ensure and maintain the maximum possible accuracy. This is to be expected, given the role of the map-maker as one contributor to an expensive production line, handling data that have been potentially difficult and time-consuming to obtain.

Contemporary developments in mapping and map-making (which are considered throughout this book) are, however, leading to subtle reassessments of the nature of accuracy and the reasons for striving for it. As more people create their own maps, as more maps whose specific purpose is uncertain are being produced, and as more spatial data are held in digital form and become capable of being combined with other digital information in unforeseen and possibly dubious ways, there is an evident need to become more knowledgeable about

Box 3.4   Contemporary mapping box – In-car navigation

The traditional applications for maps by road users are in the tasks of route planning and *en route* monitoring of progress. The road atlas and folded highway maps fulfil these roles admirably as portable and easily understandable map products. Their place is being taken, however, by computer-based contemporary route planning software, with flexibly applied route choice algorithms (e.g. shortest, fastest, cheapest routes) and up-to-date digital map databases. Such products would normally be used outside the car in preparation for the journey. Whilst *en route*, in-car navigation tools can be used.

The possibilities afforded by the combination of contemporary positioning technology, wide-area radio communications, the supply and handling of digital map data and the research and development programmes of leading motor vehicle manufacturers have led to the creation of a variety of Intelligent Vehicle Systems (IVS). A basic subdivision reveals:

- autonomous systems (with no communications, although they can receive incoming signals for positioning purposes, e.g. from GPS);
- vehicle fleet management systems (reporting *from* a vehicle to a central office and possible fixing the position of the vehicle centrally);
- advisory navigation systems (passing communications to give information *to* vehicles).

The incoming signals used by autonomous systems can emanate from land-based beacons, including those associated with developments in cellular phone networks and other wireless communication systems or from GPS (as described in Box 3.1). There may be potential problems with the latter, particularly with reception in urban areas and the likely requirement for high accuracy for road navigation. Dead reckoning systems are completely autonomous, requiring on-board sensors, such as gyroscope-based devices or, more simply, the car's own instruments (clock, odometer, speedometer, compass and differential wheel rotation).

Map-matching systems are developments of dead reckoning where automatic positional updates are implemented, at each vehicle turn for example. A digital map, structured to allow for determination of junctions and road segments, is held in the car and map-matching algorithms are used to equate the current estimate of position with actual possibilities in the data. Such algorithms are complex and manual repositioning may be necessary when returning to the road network after off-road operations (e.g. in a parking lot). Some 25% of IVS use map matching, with long-established companies such as ETAK (USA) supplying digital data and their updates from most recent topographic maps, aerial photographs and other sources.

With fleet management systems the emphasis is on the communications necessary to keep the vehicle's position known to the dispatch centre and to issue instructions. The precision of that position varies: for cross-country journeys only a rough (kilometre level) location may be required, but for city-centre fleet

(continued)

(continued)

management (e.g. for a taxicab company) an actual address may need to be integrable into the system. GPS may not be sufficient on its own for this and a series of transceivers, which transmit and receive data, may be necessary. The type of information transmitted can include basic spatial data for the area (information on road links and on transition from one link to another, e.g. forbidden turns), but also traffic updates, new recommended routes, temporary or permanent changes to the road network, weather conditions, historical data and predictive sequences for the time of travel, traffic incident detection, calculation of knock-on effects and links with emergency services.

A number of issues have to be confronted when implementing in-car navigation systems. Amongst these are the provision of spatial data in map form, the method of interfacing with the driver and the successful transmission of information to her. The display of the route and directions can be done visually but this suffers from obvious drawbacks: the attention of the driver cannot be diverted for long periods and so the image must be kept simple. Alternatively, verbal instructions can be synthesized and given to the driver.

---

the accuracy of spatial data. Extensive data manipulation, preparatory to producing maps, during their use, and at all stages in the operation of geographical information systems (see Chapter 7), may lead to confusion in 'keeping track' of data properties and the appraisal of spatial data accuracy remains fraught with difficulty.

The 'Darwinian' approach to the history of cartography (see Box 1.1) would suggest that the accuracy of maps has improved throughout history. This we know not to be true and some would contend that, given cartographers' claims that their profession is a 'scientific' discipline, the 'accuracy' and objectivity of contemporary map products leave much to be desired. One reason for this is that the perceived level of 'accuracy' of a map is, to a great extent, dependent on the task for which it is used and on the subjective assessment of the user.

The actual production of maps, charts and digital spatial data products requires a number of steps potentially from initial survey and data collection through to printing or to viewing on a computer screen. Each successive step can introduce errors or inaccuracies, due to human factors, such as incorrect measurement or faulty classification, due to limitations of the measurement and data-handling equipment, such as image resolution in remote sensing data or storage precision for digital coordinates, and due to the required generalization and limitations of the scale when representing spatial data in map form. These inaccuracies can then be compounded in the tasks of map and data use.

Components of map accuracy

Map accuracy is a reflection of how closely the view portrayed by the map represents some particular view of the real world. It covers a number of areas including the following:

Box 3.5   Contemporary mapping box – Digital data in the cockpit

The first air charts were annotated topographic land maps but they are more similar to hydrographic charts in as much as they have specific purposes: navigation, route planning and traffic control. Different phases of the journey require different charts with varying content and scales – *en route* charts are distinct from airport terminal approach charts. The type of aircraft and the sophistication of its guidance and navigation system will also determine the amount of information that the pilot needs from his charts. Military pilots, in particular, rely less on information from control towers and more on their own judgement and vision and a range of terrain information. For them, Visual Flight Rules (VFR) paper charts and their digital equivalent are used. Civilian pilots tend to use Instrument Flight Rules (IFR) data.

VFR digital data are supplied to the pilots of most modern military aircraft, along with supplementary information. Increasingly, there is use of head-mounted displays, data projected onto the visor or cockpit window and the overlay of map data with imagery (which may itself be generated from computer, radar and remote sensing sources) or the landscape outside. A list of some of the data sets available is given in Box 5.3.

The digital databases used are exemplified by radar-determined surface reflectivity in the DFAD (digital feature attribute database) which holds complete surface coded coverage of an area, incorporating 13 categories, each with a different radar signature. This is obtained from air photos, mapping and documentary information. Thus a zone or pixel (picture element – one of a number of small, regular grid-cells exhaustively covering an area) may record its percentage building coverage, percentage tree cover or some other dominant index, allowing the creation of a complete simulated radar picture. This is supplemented by vector data. Level I DFAD is for display at 1:250 000 scale; Level II DFAD is for display at 1:50 000.

The DTED (digital terrain elevation database) consists of a variety of relief features (contours, spot heights, uniform height areas, e.g. lakes, and vertical faces, e.g. cliffs) which are all interpolated to give a 100 m resolution grid DTM for Level I DTED, 30 m resolution for Level II DTED. Together, DFAD and DTED can produce a Digital Landmass Simulation System (DLMS). This product is compatible with display devices in the Tornado and other modern strike aircraft and forms the basis of pilot training with simulators.

The positioning and navigational aids on board such aircraft are integrated with the display of such data. It is clear that the traditional chart itself is liable to be of less importance in future as reliance is placed on contemporary navigation technology and image-based displays.

- How current are the data?
- How complete are the data?
- What level of generalization of the real world has been applied to the data?
- How 'correctly' have data been classified?
- How far from its true location has an object been placed on the map?

Such issues can be divided into those which address the attributes of location on the map, and those which pertain to the recorded and plotted position of the locations. The latter has been the subject of the most rigorous study, because it is straightforward to try to quantify the positional accuracy. The former is probably the more important, but is not so readily quantifiable.

Positional accuracy

Accuracy describes how closely a measurement corresponds to the truth. Unfortunately we can never know what the totally truthful measure of position is, or indeed what it means: each person's view (including that of the cartographer) of the 'truth' is conditioned by the purpose for which mapping is undertaken, amongst other factors. As we do not therefore know what the 'truth' is, all accuracy measures are relative. The term 'absolute accuracy', however, is often used to identify how well the position matches certain predetermined map accuracy standards. These rely on the fact that if a measurement is repeated often enough, then the mean of the observations is held to correspond to the 'truth'. The standard deviation of the observations (or root mean standard error, r.m.s.e.) is another valuable measure in accuracy determination as it gives an indication of the spread of the observations. The r.m.s.e. is an appropriate statistical measure when considering measurements such as those in the field, or from maps, which, when taken repeatedly and plotted, exhibit a normal distribution (reflected by a 'bell-shaped' curve symmetrical about the mean value). Some 95% of the values of such measurements will be within approximately two standard errors either side of the mean. Accuracy relates to the quality of the result, its 'precision' (the range of a series of repeated measurements, thus directly related to r.m.s.e.), and is also dependent on the 'resolution' (the minimum possible observable difference between adjacent measurements).

A quantitative statement of positional accuracy will thus include statistical measures of uncertainty and variation, as well as how and when the information was collected. Positional accuracy is dependent on the methods of data capture that do not relate to the scale of the display, whether map or computer screen. The accuracy statement is often, therefore, stated in ground measurement units. But positional accuracy is also affected by the stages of data manipulation which occur at map scale immediately after the data capture is undertaken and it may be appropriate to give the conventional description of positional accuracy on a map as a single estimate of root mean standard error of position in map units.

Such measures can be estimated because each of the stages involved in map-making can have an error value associated with it, and statistical techniques of error propagation give a combined accuracy value for the positions depicted on the map or in the digital data set. Consider, for example, the stages involved in creating a digital data set of large-scale spatial information. The first of these, the initial framework for the map obtained from a control survey, would have an accuracy, depending on the location, age and method of survey of ±0.005 mm r.m.s.e. *at map scale* (i.e. virtually zero error). Such a high accuracy reflects the care taken in observing control in the field. Further steps involve the plotting of control (which may introduce errors from the equipment used to plot maps, say ±0.10 mm), detail survey (field or aerial survey techniques, say ±0.25 mm), compilation (the bringing together of data from a potentially wide variety of other sources, say ±0.30 mm), any human input in drawing (±0.20 mm), conventional reprographics techniques (±0.30 mm) and finally data conversion or 'digitizing' methods to create a digital data set from conventional maps (±0.20 mm). Using the error propagation formula, whereby the r.m.s.e. of a set of sequential stages is the square root of the sum of the individual r.m.s.e. values squared, the positional error for these data would be ±0.577 mm: equivalent to ±11.54 m on the ground for data represented at 1:20 000 scale. This translates to a statement that 95% of the points in this data set would be positionally accurate to within approximately ±23 m (±11.54 m multiplied by two) of their true location.

It is possible to attach accuracy statements such as this to topographic map products. The United States Geological Survey (USGS) wording is that for horizontal accuracy on maps at

> publication scales larger than 1:20,000, not more than 10 percent of the points tested shall be in error by more than 1/30 inch, measured on the publication scale; for maps on publication scales of 1:20,000 or smaller, 1/50 inch. These limits of accuracy shall apply in all cases to positions of well-defined points only, [which] are those that are easily visible or recoverable on the ground.

Similar accuracy standards exist for vertical measurements portrayed in the form of spot heights and contours.

Standards exist to be tested and map accuracy standards are no exception. For topographic mapping, planimetric positions of test points are established by field teams using sophisticated, 'higher order' surveying techniques or by office personnel using photogrammetric methods to determine positions from aerial photographs. Vertical tests can be run separately to determine precise elevations. For the USGS, the wording of its National Map Accuracy Standards, which 'guarantee' accuracy, cannot be printed on a map which fails to meet certain thresholds.

Although data accuracy at capture stage is scale-independent, scale is important in the sense that the resolution of a paper map sheet is limited, which is why positional accuracy of data on a 1:10 000 paper map will be different from the same data rendered at 1:100 000. Such variation in accuracy is, however, the result of the limited resolution of the smaller-scale map and the generalization needed to render the data at smaller scale. The positional accuracy of the raw data remains as it was when captured.

Positional accuracy should also be assumed to cover the interrelationships of

Table 3.1   A confusion matrix indicating the land cover class of locations as represented on the map against their land cover as determined by ground truthing

Class on map	Class on ground				Total sampled sites
	Forest	*Pasture*	*Arable*	*Bushland*	
Forest	93	8	15	—	116
Pasture	6	65	23	1	95
Arable	11	34	503	32	580
Bushland	5	—	21	72	98
Total sampled sites	115	107	562	105	889

elements on the map face. Such factors as ensuring that rivers cut through contours at the appropriate place and that railways do not appear to run in the sea are not quantifiable but reflect the accuracy of the map. Current efforts by map production agencies to establish and abide by quality control schemes which address these concerns are increasing awareness of the importance of map accuracy among both map-makers and map users.

Representational accuracy

When the attributes (as opposed to the position) of locations are examined, the problems of accuracy may, at first sight, appear to be considerably easier to handle. It is usually straightforward to determine whether the name given to a location is incorrectly spelled on the map (although there may be alternative names or spellings in certain parts of the world). The classification of much data, particularly of a nominal (category) or ordinal (class ordered) nature can generally be either right or wrong: thus an area of deciduous woodland classified as coniferous, or a major road categorized as a minor road, would not normally involve probability statements to quantify their error. However, once such data are manipulated, combined with other data (e.g. during overlay operations in geographical information systems, described in Chapter 7), revised or reclassified within digital mapping or GIS, the propagation of error and misintepretation can occur in unpredictable ways.

Much testing is conducted *a posteriori* using techniques such as error or confusion matrices (Table 3.1) which compare the attributes of locations sampled on the map and their true nature in the field. Such testing is sometimes called 'ground truthing'. A totally accurate classification on the map would reveal all the numbers in the matrix in Table 3.1 to be on the diagonal.

One check is to give a percentage figure to the ratio of correct values to total values (i.e. sum of diagonals over total) which would yield an 82% accuracy level for the map in this case. We can also determine accuracy levels for each class; e.g. pasture was classed correctly on the map in 68% of the sites sampled, and in 24% of the cases it was incorrectly classed as arable. A measure which considers significantly unequal sample sizes (as here where arable land is clearly the dominant category) and likely probabilities of expected values for each class, is Cohen's kappa:

$$\kappa = (d-q)/(N-q)$$

where N = total number of samples; $q = \sum_{i=1}^{n} (\text{row}_i \times \text{col}_i)/N$; d = total number of cases in diagonal cells. For Table 3.1, $N = 889$, $q = 404.67$, $d = 733$ and $\kappa = 0.678$. The optimal score is 1.0.

The testing of maps by ground truthing (and also the classification of satellite remote sensing imagery which is done using similar techniques) raises questions about the methods used to select the sampling sites for testing. Sampling strategies can vary from regular, systematic coverage of data points over the whole map sheet, to a random selection of sites for assessment. An aim may be to standardize the sample size for each class, or to base sampling on pragmatic concerns such as accessibility of locations for the field team.

If data exhibit any fuzziness (e.g. the boundaries on a soil map which are only guidelines to the distribution of differing soil types and cannot be viewed as discrete and absolute locations of change), then the use of the data becomes more difficult. The handling of uncertainty of this nature has led some mapping scientists to examine 'fuzzy logic' which deals with probabilities of occurrences and their interaction. The type of answer to a map- or GIS-based query which planners may be able to obtain in the future would be of the form: 'there is an 87% probability that this area is a suitable location for growing irrigated rice'. Such statements rely on the examination of a potentially large number of spatial variables – in this case soil type, terrain data, climate statistics, workforce availability, etc. Clearly such an examination is most efficient in the digital environment using GIS. Map-based queries tend to be simpler, but it is just as important to recognize and take account of the inherent error in the graphical information also.

Accuracy information is often stored as 'metadata' (information about a map or data set held in readily available form). The accuracy standard written on USGS mapping is an example of metadata. For GIS data sets, the accuracy information is often made available with the data, and it may cover classification accuracy (e.g. Cohen's kappa), for land-use information for example, as well as other data relevant to accuracy, such as the date of data capture and the organization responsible. It is important that attention is paid to any such accuracy information before and during map use and GIS data manipulation tasks which are undertaken.

Summary

Maps (and collections of digital spatial data) are representations of one view of reality, constructed using techniques that are inherently error prone and are likely to introduce generalization. Data capture, manipulation (cartographic generalization in particular) and presentation are all likely to introduce uncertainty and inaccuracies. As indicated in Chapter 1, maps are not faithful reproductions of reality; they are filters through which an individual or societal view of reality is expressed. The information they portray should be treated with caution as, even at a technical level, no maps tell the 'truth'. The relationships among map, map-maker, map user and notions of the 'truth' are considered in the next chapter.

Further reading

An introduction to navigation which includes an assessment of the impact of GPS is found in N. Ackroyd and R. Lorimer, *Global Navigation: A GPS User's Guide*, 2nd edition (Lloyd's of London Press, London, 1994). GPS itself is described in detail in Chapter 7 in G. Seeber, *Satellite Geodesy: Foundations, Methods and Applications* (W. de Gruyter, Berlin, 1993). The construction and use of navigational charts is covered in Chapter 20 in J. S. Keates, *Cartographic Design and Production*, 2nd edition (Longman, Harlow, Essex, 1989). The work of Mercator and other cartographers in seventeenth-century Europe is described in L. A. Brown, *The Story of Maps* (Dover, New York, 1977). A comprehensive insight into all aspects of cartometry (measurement on maps) is given by D. H. Maling, *Measurement from Maps* (Pergamon, Oxford, 1989). The possibilities of using data from maps to create GIS data sets is considered in a number of chapters, notably those by Goodchild and by D. Rhind and P. Clark, in H. Mounsey (ed.), *Building Databases for Global Science* (Taylor and Francis, London, 1988). The issues of error analysis during data manipulation within GIS are addressed by many distinguished authors in M. Goodchild and S. Gopal (eds), *The Accuracy of Spatial Databases* (Taylor and Francis, London, 1989).

Chapter 4

Representing others

Introduction

There is a growing school of cartography which suggests that many of the assertions made in the last chapter are a smokescreen for the actual purposes and origins of most maps. This school argues that maps are about social control and are usually created to serve the designs of their creators rather than to inform 'the public'. Arguments about accuracy are often excuses for not showing certain things and for including others. Governments and other powerful organizations which in practice control most cartographic production choose what information they collect and how they show it in quite partisan ways. Because of this we should not take maps at their face value or see cartography as the purely technical operation of translating reality onto paper or screen. Much more can be read into maps than this.

'Representing others' is a misleading title for much of this chapter as most of the examples do *not* concern representing others but are about representing ourselves. 'Others' here refers to peoples and places which are not dominant in the media cultures that determine how difference is represented, whereas we (working in universities and having easy access to students to teach and publishers to print books, such as this) are in a relatively powerful position to contribute to debates on representation. Still more powerful have been the agencies or their sponsors which have traditionally produced maps, and yet these images are often accepted uncritically as general records of the past and present, often perhaps because of their overtly technical appearance. Almost by definition, 'others' have not been well represented on maps, but this dearth of representation appears to be beginning to change.

Not only may the way in which we have looked at and reacted to traditional maps in the past have been naive, but there are many different types of map representations other than those commonly thought of as maps. These are referred to here as map-like objects (as defined in the Introduction). For instance, people's mental maps of an area can produce fascinating pictures. Tourist, commercial and stylized maps are also often thought not to be 'the real thing'.

Mapping and map-making are breaking out all over the place, from the Internet to the primary school. It can be argued that, just as writing was once controlled by those who could afford scribes or who could (in Europe) read Latin, cartography has been the preserve of the powerful who had an interest in maps being produced and in deciding what symbols were placed on them. These were people who could afford to employ the cartographers – a group of artisans who until quite recently were very expensive to employ – to draw maps. Mimicking the growth of literacy and writing, which has been escalating for several hundred years, map-making is also becoming less exclusive, and what, who and where is mapped is now much more extensive. This chapter is about the ways in which certain scholars are coming to understand the old cartography, and how it is being transformed by the new.

Who are maps made for?

Maps are like clocks – once we did not need them. Then, as people's lives became more interdependent (some would say servile), accurate timing started to matter more and more. There are few people today who do not wear a watch on their wrist to tell them 'when' they are. Only a few generations ago people managed to get through their everyday lives without the need, and often without the means, to know the time. Then factories were built and their gates opened at a certain hour, the post had to be delivered at a certain time, trains ran to a time-table: time became more important. So too with the map. Few cars will be found without a road atlas or A–Z street plan in the glove compartment; few diaries without a train or underground map in them; few city centres are now not adorned with maps for the tourist and other visitors. The spread of mapping into many people's lives began when it was realized that land could be farmed more efficiently if it was measured accurately, and just as the introduction of factories required clocks, the enclosure of agricultural land needed maps. The purpose of these maps was not just to measure the land but to legitimize the transfer of its ownership. You cannot usually own what you cannot see. Maps were, of course, also drawn for use in war, but it was in peacetime that their importance increased most dramatically over the last 400 years. The new canal, road and rail infrastructure of countries could not be built without accurate maps of the terrain. The spoils of lands which were new to western eyes could not have been subdivided so neatly without the surveyor and the cartographer. Just as clocks tell you what time is yours and what time belongs to your employer, maps strengthen the boundary lines between properties. When the state paid for maps to be drawn it usually did not do this because there was a popular clamour for these pictures, but because in order to rule it needed to know what it was ruling. Maps are also required by people who make money and, put crudely, the state needs people to make money – so that it can tax them (all these issues are explored further in the next chapter). It is also the case that people now travel more and further than they used to, just as they now tend to engage in a more varied range of activities in a day. The spread of maps and clocks was necessary to change the world and they have both become essentials for most of us as a result of those changes.

Cartography and capitalism

The confusion between the benevolence and self-interest of those who fund the production of maps continues to this day. For instance, when the United States government allows people around the world to use the hugely expensive satellite-operated Global Positioning System that it funds, it may not be doing this entirely out of goodwill (see Box 3.1). You have to ask yourself who benefits from the sales of the gadgets that are needed to use this system (firms in rich countries?), and who needs this level of geographical 'accuracy' (industries prospecting for oil or cutting down forests, perhaps?). It is also important to remember the origins of systems such as these. The United States military complex and its missiles only know precisely where their enemies, and they themselves, are because of the global distribution of the signals from these satellites. The state provision of accurate time-keeping by electronic means was one precursor to GPS systems designed to allow accurate 'place keeping'. Only a few generations ago the railway station at Bristol, England, ran clocks that were seven minutes later than those in London (to allow for the movement of the earth under the sun). Just as radio is used by GPS, radio signals allowed the first accurate nationwide time-keeping – provided by the state. Today, computers in workplaces across the world constantly check each other's clocks to achieve the extremely high levels of accuracy required for the efficient operation of communications between them, without the use of radio waves (the operators in charge of the most accurate machines are known as time-lords!). It is likely that maps will be constructed and updated in a similar, if more complex, way between computers in the future. Although this has certain benefits, it will propagate the myth of there being one true map of the world; just as the use of Greenwich Mean Time in Britain means that people forget that the sun is no longer at its zenith in Bristol at noon. Britain may soon move to standard European time so that businesses are not inconvenienced by one hour's difference in office working hours. Maps too often reflect the interests of those whose interests are to make profit.

The links between capitalism and mapping are sometimes very obvious. The Rand McNally corporation in the United States of America printed its first custom-made maps for local merchants in the 1920s (Akerman, 1993). The location and names of the merchants' businesses were printed on the maps for a fee and these maps were then sold on to the car travelling public at low cost, or often given away free of charge. The sponsorship of other American road maps by national newspapers and oil companies has continued this tradition and provided cheap maps for millions of people. However, on the flip-side these maps have a cost. They are blatantly commercial, cajoling drivers to travel along the route that will make the map supplier most money as the travellers stop off at the sponsoring organization's outlets. More sinister than this, however, is the subtle selling of a way of life through such maps. Akerman suggests that the Rand McNally corporation not only sold road maps, but sold the idea and landscape of automobile America – the right to travel freely and easily and at the expense of what was there before the roads and the people came. Most of these 'free' maps had pictures on the cover of the American nuclear family (invariably white and middle class, with the husband driving and the wife 'passive' in the 'passenger seat'), enjoying their private, luxurious form of transport, isolated from the countryside through which they travelled, sampling the unthreatening wilderness (tamed by the mapping process) which was regarded as their inheritance.

The true depth of the integration of cartography and capitalism is rarely revealed by such simple examples as that provided by Rand McNally. Increasingly, writers are arguing that most maps are not made to help people find where they are going, but to show the lie, ownership and management of the land for the businesses and the governments that are thinking of selling there, or working there. Maps, in practice, tell the ordinary traveller more of where they cannot go – where they have no right-of-way – than of where they can travel. One defence of this habit can be mustered by looking in more detail at who pays for maps to be made. The full cost is certainly not borne by the public. In most countries, the production of paper maps is still largely subsidized by government. However, the creation of digital maps is mostly the work of the utilities (companies and nationalized industries which provide gas, electricity, water, telephone lines and so on) and private business, and there are already many, many more digital maps in the world than paper ones. This is despite the fact that most digital maps are temporary, existing on disc or in the memory of a computer for short periods only (if we were to include these also the dominance of digital maps over paper would appear total). Why should business be investing so much in mapping? Quite simply, as illustrated above, there is money to be made out of space. In the past, the exploitation of new markets was achieved largely through human travel and exploration. Nowadays, markets are segmented and customers targeted more efficiently by 'analysis', using tools such as geographical information systems (see Chapter 7) and a key to their segmentation is geography. From 'red-lining' out areas not to trade with, to locating 'hot-spots' of affluent buyers, maps make people money.

Political cartography

Maps can also be seen to operate at a more subtle level again, beneath that of tools for the exploitation of space. The fact that governments, and the wider operations of the state, support their creation suggests that they also have a political role to play. Maps present an image that is 'alternately denying and celebrating the "patchwork country" existing within the boundaries of each political state' (Rundstrom, 1993, p. vii; see Further reading). The outline of each country is a potent political symbol. Only rarely is this completely a physical feature such as the coastline. It is usually a political artefact. Not only do boundaries give the country a shape, but they suggest a uniformity within that shape which separates it from the outside, from what is alien or foreign. But maps which show a patchwork of diversity within the boundaries of a country also serve a political purpose. They can be used to say: 'look at this mess, only by working together, only by strong government, can we make this work'. Political atlases which celebrate the diversity of the state can also be used to legitimize its existence. A clear example of this, which is evident even in its title, is the beautifully produced *We The People: An Atlas of American Ethnic Diversity*

(Allen and Turner, 1988). This atlas is a celebration of ethnic diversity, but even more it is a celebration of being 'American'. One critism of it could be that it helps to sustain the myth of the United States of America as a 'melting pot' of people from many different backgrounds. An atlas entitled 'We In Power: An Atlas of American Ethnic Divisions', might use much the same information but would present a very different picture of that country. Schoolchildren and college students are shown maps and atlases not only to educate them but also to help give them a sense of identity of the country they come from. Maps can be powerful political symbols.

Another use of maps is far more overt: this is to assimilate people living in a particular place to another group's way of life, by mapping them as if they were no different. There are several well-documented contemporary examples of this in relation to the United States of America and its indigenous peoples (see Box 4.1). At another extreme, maps can be used not only to make whole peoples disappear (through assimilation), but literally to exhibit them. Some American tourists' maps do this and could be said to to have aided, particularly in recent years, the destruction of Indian ways of life by leading thousands of tourists to their homes and sacred sites. Figure 4.1 shows the legend of an official (rather than tourist) map of Indian lands, but note that the key symbols are the positions of reservations, wildlife refuges, tourists' complexes, monuments and the interstate highways needed to visit all these curiosities. Maps of sacred Indian sites are now being used by tourists with an ecological bent who may be causing as much damage with their feet (and their maps and their money) as their ancestors achieved with guns and disease (and, again, their maps).

The concern of indigenous peoples about the making of maps showing sacred sites is also shown by Aborigines in Australia. The transfer of 'Dreamtime' tracks, important linear features in the barren landscapes of the outback, onto topographical maps is regarded by them as breaking secrets, to which the uninitiated (particularly exploration geologists and politicians) should not have access. Aboriginal maps are only readable by the initiated; they, themselves, wish to maintain the secrecy of certain categories of knowledge which they preserve on map-like objects. Maps of areas such as the Australian desert are often not compiled by, and in the interests of, the people living on the lands they show. For instance, one contemporary use of these maps is to mark off the territory that was irradiated during the extensive testing of British nuclear weapons in the outback. Before the bombs were tested the surveyors came to map the land. Their maps were a small part of the 'scientific' decorations that were needed to help those conducting the tests feel they were involved in a legitimate and necessary activity. What happened in the outback of Australia occurred only a generation ago and was instigated by the British Army. Below we discuss more historic mapping by that organization, but it is important to realize that this is not all about the past: it is just that we know more about the past – contemporary mapping by such organizations is, of course, highly secretive. The use of maps to control territory, an essential role of government, is examined in the next chapter.

Box 4.1   Contemporary mapping box – American Indians and geographical information systems

Robert Rundstrom (see Further reading) has collected a number of examples of how geographical information systems are being used to assimilate the remaining indigenous peoples of the United States of America into a 'white way of living'.

The first example is of the Zuni Indians who live in pueblos (small towns and villages) on a reserve in western New Mexico. As part of a GIS-related health project, a digital gazetteer 'had' to be constructed of all addresses in McKinley County. This county contains most of the reservation. The work was being done by government officials and university researchers. Unfortunately for the researchers, the Indians tended to use neither house numbers nor street names. Undeterred, the team decided to give their roads English names and their houses street numbers in a 'rural addressing project'. In a few cases Zuni names were used but only if they were capable of renditioning in English and not 'declared too "humorous" for a proper road name' (Rundstrom, 1993, p. 22). Thus, a bit of humour and a great deal of history was being written out of the landscape of McKinley County, New Mexico, by the new computer cartography. The Indians have had their health care delivered slightly more efficiently, but have lost their heritage as they have been assimilated into the digitizable society encapsulated within the GIS.

The second example tells a similar story and shows how such practices can be endemic. There is a Hopi Indian reserve in Arizona which the National Mapping Division of the United States Geological Survey chose to survey for place-names to update topographic maps. The surveyors asked the Indians for their names for places in the reserve and promised them that these would be used on the updated maps. Again, however, the Indian way of life would not fit into the GIS or into the conventions of modern official American cartography. Hopi place-names often contain numerous accented characters when written in the Roman alphabet and can be very long. Government cartographers objected to the length of these names because they were hard to fit on the map and because the small diacritic marks might look like other cartographic symbols. What is more, it was not easy to store these place-names in the GIS system being used to do the mapping. Just as the use of Indian languages in schools was first attacked overtly in the last century, the tradition of cultural assimilation of indigenous Americans continues, albeit more subtly, today.

Parallels to these two American cases can be drawn from around the world. In Wales, for instance, the British government, through its schools, tried (in the nineteenth and early twentieth centuries) to outlaw the speaking of Welsh. It is only in recent years that the Welsh language has been used on road signs (in tandem with, rather than replacing, the English). Welsh is now taught in schools and it is also used on maps (despite the cartographically upsetting length of some Welsh place-names). However, the assimilators may have already won in Wales, for although there is currently a rise in the use of the Welsh

(continued)

(continued)

language, this is largely amongst English immigrants to that country, and the English are moving in droves to retire to rural Wales where the 'quaint' place-names are now being preserved. Thus what is happening in the United States of America has happened before elsewhere. No doubt the Hopi place-names will appear on Arizona's maps sometime in the future, long after most of what is left of the Hopi have stopped using them, but in time for the retirement migration of wealthy Americans. Rural Welsh valleys used to be as poor as Indian reservations and they too were mapped by outsiders, not primarily for the benefit of those who lived in the valleys, but for the people from England who governed them.

Map compilation

It was suggested in Chapter 1 that maps are not compiled, and hence are not found, in every human society. To appreciate how the compilation of maps affects our representation of others it is helpful to first ask why some people draw maps and others do not. There is an argument, which is often repeated, even in modern histories of cartography, that certain people cannot draw maps (Lewis, 1987). Another view is that certain peoples did not need to draw maps because their societies were too small to warrant them. The received wisdom is that 'smaller, less developed societies have no need to map land ownership, tax assessment districts, the topography of tank attacks, sub-surface geology likely to contain oil, sewer lines, crime statistics, congressional districts, or any of the rest of the things we find ourselves compelled to map.' This is, of course, an ironic quote, drawn from Wood (1992, pp. 5–6), who goes on to say that we should note, however, that 'this does not mean that they don't create in their heads dense multi-layered, fact-filled maps of the worlds they live in'.

When map compilation of many different groups of people is examined in detail, cartographers can arrive at an uncomfortable conclusion: '[it is] difficult to hold unquestioningly the belief that there are certain objects called maps, which most, if not all, human groups use to communicate information about the location of geographical features' (Orlove, 1993, p. 29). The conventional view is that a picture is a map when it resembles the world in miniature and the 'better' it resembles the world (i.e. the more accurate it is), the better a map it is. Denis Wood has argued differently: that the authority of a map is not derived from its accuracy, but from the authority of the person who draws it. A picture is a map when it is drawn by someone with the authority to draw maps (Wood, 1993).

Compiling maps of the colonies

A classic example of a group with the authority to draw maps was the British Army under the auspices of the British East India Company. The traditional view of western topographic map compilation, the epitome of which was the

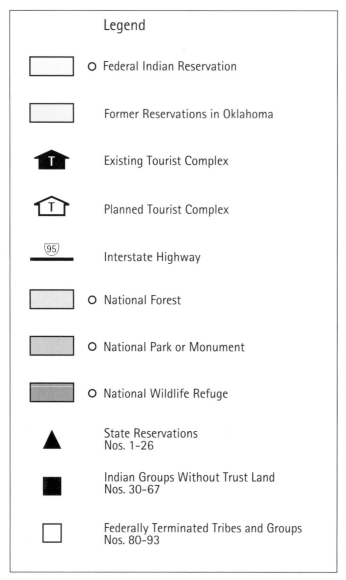

Figure 4.1　Legend of the Bureau of Indian Affairs' map of Indian Land Areas (from Cole, 1993).

survey of the Indian subcontinent, has been re-examined by cultural and social cartographers and found wanting. For instance, the story of the triangulation and mapping of India by the British East India Company's surveyors suggests that a precise trigonometric survey was first made and that the picture of the land in between the survey was subsequently filled in so that 'the detailed work of many different individuals [can] be brought together without having to be "adjusted" to fit one another' (Edney, 1993, p. 61). Edney has found that the surveying of India was in fact a far more *ad hoc* process than this, with the precise

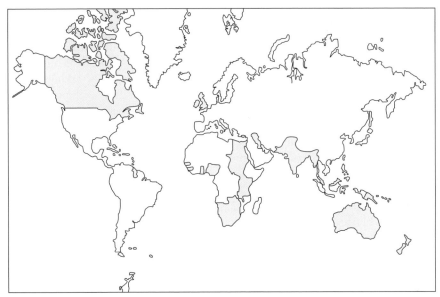

Figure 4.2   Map of the greatest extent of the British Empire drawn on a Mercator projection.

triangulation, which was supposed to be done first, often being left to last because of its complexity and expense.

So what was the purpose of the triangulation of India if it was not to provide a basis upon which more detailed surveying could be later conducted? Edney offers two possibilities. The first was to set Europeans apart from the Indians (meaning here the indigenous peoples of India) to show that Europeans could do something that Indians could not do – that they were different and that they were born to rule. The second possibility was that the triangulation provided an image of the British Empire as a single 'imperial space', as an entity which had a reason to exist. Triangulation provided the best means of knitting the disparate lands of the Empire together in people's minds. When George Everest completed the 1500 mile triangulation of Lambton's Great Arc from Cape Comorin to Dehra Dun he was doing much more than precisely demarcating the interior of a subcontinent: he was tying two far-flung parts of the British Empire together, and to the rest of the Empire, with 'science' (Figure 4.2).

The discipline of geography has had a long association with colonialization and empire building, but it is possible to argue that cartography has had an even deeper involvement than this. The writer who has opened up this field more than any other, Brian Harley, saw maps as weapons of imperialism which were at least the equal of guns and warships (see Box 4.2). They were used to claim lands before they had even been occupied and before geographers knew at all precisely where those lands were (see pp. 88–89). Surveyors worked with soldiers to produce maps which pacified the colonies and facilitated their exploitation. From the Greek to the Roman and then Spanish and Portuguese empires, maps were the symbol used to legitimize conquest. This process became most prolific in the nineteenth century with the scramble for Africa

Box 4.2   Personality box – Brian Harley: taking mapping apart

Brian Harley changed the way people think about maps. During the 1970s cartographers developed techniques to try to understand maps, such as models of cartographic communication, which involved interrogation – taking 'maps' apart (see Chapter 9). Harley moved this emphasis onwards to the interrogation of the whole cartographic process – taking 'mapping' apart to try to understand maps and how they represented the world. To do this he looked at the social relations behind map production, the working practices of the cartographer, the politics of power and the surveillance of the state. He argued that 'the scientific and objective character of maps is simply an illusion . . . created in the 18th century and fuelled by our increasingly technological guild of cartographic practitioners' (Edney, 1992, p. 177).

If Brian Harley were alive today he might be amused, or maybe a little perturbed, at how much information is available about him in digital form: by the end of 1995, 100 of his separate publications had been cited in learned journals referenced on the Internet. He began writing in 1958, two years before completing his PhD in historical geography, but his most widely read work was produced in the last ten years of his life. His essay on 'Cartography, Ethics and Social Theory' had the greatest impact, and was published in *Cartographica* (Volume 27) in 1990, a year after the publication of 'Deconstructing the Map' in the same journal. Most cited, however, was 'Maps, Knowledge and Power', his chapter in Cosgrove and Daniels (1988).

Some of the points Harley made were very simple. For instance, just by looking at the orientation of a map you can guess who made it. British world maps put Britain in the middle, not for 'scientific' reasons but to demonstrate British supremacy. One reason for cartographers' mass hysteria surrounding the publication of the Peters projection (see Chapter 1) was not its lack of scientific rigour or any plagiarism associated with the product, but the breaking of a geopolitical code of how the world ought to be shown. However, equally important to Harley was what was not shown on maps – which places were deemed not to exist, which buildings, bridges and sites were secret. Map-making gives people the means to present space as something that can be manipulated, often at the expense of those who live there.

Despite his large academic output and the widespread influence it has had on other researchers and students, Brian Harley's greatest impact may have been outside academic life, as he wrote and lectured extensively for amateur historians and the public at large. His *Ordnance Survey Maps: A Descriptive Manual* (1975) was named the outstanding reference book of the year by the British Library Association. Far more people read books and pamphlets such as this than will consult the multi-volume *History of Cartography* (which he co-edited with David Woodward) or read articles in scholarly journals. Harley wrote tirelessly on many fronts. Many of those who wrote his obituaries put this down to his modest background. Born in Bristol in 1932, but brought up by a foster family in Staffordshire, he won

(continued)

(continued)

a place in a grammar school in Wolverhampton and this allowed him to go to university.

In 1952, following national service, John Brian Harley arrived to study for a degree in geography at the University of Birmingham. He stayed on at Birmingham to undertake his PhD and, after a very short spell of school teaching, took up an assistant lectureship in Liverpool in 1958. He resigned from this post in 1969 following lack of promotion and spent a short time in publishing, most notably helping to start what became the long-running Cambridge University Press Historical Geography series. A year later he was appointed to a lectureship at the University of Exeter, which was converted to a readership in 1972. Then again, in 1986, partly due to lack of promotion, he resigned, to take up a full professorship in geography at the University of Wisconsin-Milwaukee. The fact that the value of his work was only fully realized late in his career, particularly outside his subdisciplines of cartography and historical geography, is testimony to how he was changing the way that people now think about maps, rather than producing orthodox scholarship that would have given him an easier career.

Brian Harley died of a heart attack while driving to work in December 1991.

which was dissected into European territories by straight lines drawn on the 'ultimate arbitrator', the map.

Interpreting the interpreters

> Whether a map is produced under the banner of cartographic science – as most official maps have been – or whether it is an overt propaganda exercise, it cannot escape involvement in the processes by which power is deployed. Some of the practical implications of maps may also fall into the category of what Foucault has defined as acts of 'surveillance' notably those connected with warfare, political propaganda, boundary making, or the preservation of law and order. (Harley, 1988, p. 279)

Many maps have been held and produced in secret. Examples can be drawn from as long ago as the sixteenth century, with the policies of the Spanish and Portuguese governments. The practices of the Dutch East Indies company continued this tradition, as does the modern-day commercial secrecy of the British Ordnance Survey (the British public is encouraged only to 'lease' digital copies of Ordnance Survey maps – at huge expense; see Chapter 7). However, there are more subtle forms of secrecy and these are often used. Harley defines cartographic censorship as the removal of map features which, other things being equal, we might expect to find on a map. Harley provides many military examples of this, from deliberately not updating maps to deceive enemies (as Frederick the Great ordered), to the relocation of towns on 1960s Russian maps to confuse American military planners who were attempting to target nuclear missiles to land on those towns (Harley, 1988, p. 289).

Interpreting map content and appearance

More subtle again are 'distortions' in maps which are not deliberate. It is quite
easy to look back at ancient maps, for example the *mappae mundi* discussed in
Chapter 1, and see how their design often reflected the beliefs of those times; for
instance, that the universe revolved around Jerusalem. However, it is doubtful
that the cartographers of that time deliberately set out to deceive people: many
almost certainly believed that they were drawing the truth. Similarly, the widely
used Mercator projection, which enlarges Europe and shrinks the equatorial land
masses whose people Europeans later came to govern, was not a deliberate distortion.

Subtle, often thought of as unconscious, distortions can be seen as 'silences'
on maps. Old maps of the countryside often excluded poor farm workers'
cottages because they did not fit into the idyllic rural landscape the maps were
drawn to portray, or because they appeared to visually trespass on a rich
landowner's property. Similarly, Victorian maps of towns often neglected to
include the courts and rookeries where the urban poor lived, instead
emphasizing the more affluent streets and squares. The current practice of using
equal-area maps to portray census data continues this tradition by making the
areas of cities where the poorest people live too small to see (see Box 8.3 on cartograms).

Apparently innocuous map symbols can also convey subliminal messages to
the map reader. Take the hierarchical signs used to represent towns and cities of
varying populations. When placed on a map these symbols present an image
which often reinforces social–spatial hierarchies such as the organization of
government or the church, making this order of power appear to be natural,
almost God-given. Maps can give shape to abstract notions of who has power
over what. They thus tend to present the status quo as given, rather than as an
option and they are often used as tools by those who are conservative to change.
In general, maps are only changed years after social systems have altered. They
present a picture of the past as 'correct' – a picture against which current changes
appear as errors.

But how do map-makers get away with this and is cartography really so
Machiavellian? They get away with it largely because they do not know they are
doing it – they are simply continuing conventions: there often is little else practising cartographers can do. Any kind of real-world depiction, whether it is a
map, a story or even a photograph, contains unconscious messages as well as
conscious ones from its author. There can be no neutral map of a place and certainly no neutral map of the world. One cartographer's politically correct town
map may be another's vision of a totalitarian society. However, what does make
maps different from other types of representation is that they are so fact-filled.
There is no other way of cramming so much readable information onto one
piece of paper than by drawing a map. And, because maps are so full of facts it
is easy to forget that their signs are also symbols, that the drawing is not entirely
constrained by scientific rigour, and that there are an infinite number of pictures
to be drawn of any place.

An infinity of images

How though can an infinite number of pictures be drawn? As this text is being written, the way in which maps are presented is changing more quickly than ever before. The engine of this change is the same as that which has always changed cartography: new technology (within the context of a changing society). However, this change is a little different from its counterparts in cartographic history. Improvements in surveying and navigation technology and in drafting and printing techniques increased the extent and content of mapping by, at most, an order of magnitude with each innovation. In the past, innovations in mapping technology tended not to occur at the same time as increases in the amount of spatial data available to be converted into maps, or simultaneously with radical change in prevailing world-views. Today, however, we are in the midst of a veritable revolution both in changing interpretations of the world (along with the availability of data to support them) and in the technology at hand to render them as maps. Today's new technology is therefore currently having a far greater impact than previous leaps in mapping and will have a far greater impact still if it continues to evolve at its current rate. One major component of this new technology is the 'Internet' which allows almost anyone to 'print' (visualize) a map and almost anyone else to view it. The Internet has become famous not because it permits computer files to be transferred with ease around the world, but because it allows the real-time transfer of full-colour images. What is most important about these images is that a practical infinity of them can now be drawn, stored and retrieved. The problem initially identified from historical cartography, that no map is an accurate impartial record, has therefore been addressed with a solution from the technological side of the field – a mechanism whereby everyone can document their partially accurate view of the world in a map that anyone else can view. The Internet is, however, not the anarchic Utopia which it was often portrayed to be in its early years (we return again to the limitations and possibilities of such new technology in Chapter 9).

Unlike paper products, electronic images can be animated to show a sequence of events or to allow the viewer to zoom and pan across them. Figure 4.3 shows an inset of an interactive map of housing in Buffalo where the author has zoomed in from an image of upstate New York, to a scale at which he can look at his own house. Maps on the Internet can be made to be interactive so that a 'click' on the map reveals more information about a certain place than can be shown on the initial screen. The equipment needed to produce these maps and to sign up to the Internet can now be fitted to the most humble of microcomputers. Schoolchildren, pensioners and even the most technologically illiterate can make it work. It is not difficult. The technology of big brother, the networked computer which make it possible to record and update a billion names and Zipcodes or to instantaneously track millions of credit card financial transfers is now being used by ordinary people, to present their representations of the world to others. At the forefront of these processes are researchers, who are now able to present their results to the world and to explore more easily other scientists' experiments and results. And they do this at the same time as these results are

Figure 4.3 An inset, portraying the map author's house, of an electronic map of 250000 dwellings around Buffalo (from Batty, 1995).

revealed to the select few attending an academic conference (see Box 4.3). The Internet allows people to draw and present maps to a huge audience, without ever having to print them or having to have them accepted for publication by powerful cliques often driven by particular interests. Truly power to the people?

If the above sounds a little naive that may well be because it is. In actual fact only a tiny proportion of the world's population have access to the Internet and fewer still have the resources to place detailed representations of the world on-line. Proponents of the technology argue that as the hardware and software become cheaper it will diffuse to become available to even more people, living in more and more isolated places. But did this happen when other new technologies were introduced? Why is cheap paper printing not universally available? Why cannot just about anyone publish a book of maps? Is it, perhaps, more likely that, as the Internet develops, sections of it will be cordoned off to outsiders, maybe through pricing it at a high level, maybe by citing excuses such as 'academic integrity' or that 'the technology cannot cope with all the traffic'? It is unlikely that those with most power in this world will be happy to give up some of it, just because it has become slightly easier to do so.

Summary

From the parchments of a millennium ago, to the digital animations you can view in a few seconds from now (assuming you are reading this near a networked computer!), maps have always been about knowledge and power – selecting

Box 4.3 Contemporary mapping box – Cartography and the Internet

Cartography burst upon the Internet in the early 1990s. To find examples of such cartography you only have to connect to the World Wide Web (part of the Internet) and type 'cartography' into a 'search-engine' and see what appears. A search-engine is a piece of software that sends 'search-robots' around the world-wide computer network. These virtual robots collect pointers to keywords on their travels, hence when you ask for 'cartography' the software will point you to any of the sites its robots have visited where that word appears in a 'document'. In one sense the World Wide Web is the largest map ever drawn. It is also the map drawn by the largest ever number of people. It is a map so vast and changing so quickly, that even tireless electronic travellers (the search-robots) can never get to the end of it!

The search-engines will come up with documents put on the web by a variety of organizations and individuals, from national mapping agencies to commercial companies, from research scientists to students in higher education. The type of map that appears when such pages of information are accessed range from simple location maps, perhaps created using desk-top mapping software packages, to extracts from traditionally published paper mapping that has been scanned into the host computer. These maps are published in this way as images to inform, to sell paper maps, to illustrate hypotheses and research findings, to demonstrate availability of data sets, to advertise, and to flaunt expertise and knowledge. There seems to be no limit to the abilities a 'connected' individual has to present their world-view to the mass web audience in a way that could never have been envisaged by practising cartographers of even five years ago.

One of the first academic papers with extensive links to cartography to appear on the web was Michael Batty's (1995) 'The Computable City' which begins: 'By the year 2050, everything around us will be some form of computer.' Batty's paper contained references to numerous examples of mapping on the web. These included a map of San Diego road traffic volumes which was updated every 15 minutes so that drivers could see where the congestion in San Diego was; the entire 1990 United States census which is available on-line from the Lawrence Livermore Laboratories of the University of California to anyone in the world; and an animation of 170 years' worth of urban sprawl around the San Francisco Bay area which had been put together at a site in Silicon Valley. The World Wide Web pointer to Batty's paper was http://www.geog. buffalo.edu/ although it is unlikely to still be there by the time you read this (one reason being that Michael Batty left the United States of America to take up a professorship in London shortly after establishing the web site – the Internet is almost as fluid as people's lives!). That fluidity, in addition to being useful and necessary, is a major problem. Items can come and go before even the search-robots have time to find them. It is ironic that the paper copies of this early paper by Batty (1995) on the Internet and cartography may well outlive the electronic record, despite the initial widespread inaccessibility

(continued)

(continued)

of such paper products (for instance, getting hold of the paper copy may be possible by inter-library loan).

There is another problem with mapping on the Internet which the examples given above typify. Just as much traditional mapping can be viewed as Eurocentric, having a 'First World' bias, so cartography on the Internet currently tends to portray only a small portion of the globe in detail and only those details of interest to a small number of people. The technologically literate of California currently dominate the production of information on the web. The examples above were not drawn from (and of) San Diego, Berkeley and San Francisco by accident. The web is essentially American – people communicate on it in English, and its origins, from the Internet, are military. Where, for instance, can you find maps of traffic volumes in Mexico City; where can you access an on-line version the Union of Myanmar's latest census; or see an animation of the growth of South African townships? Whether this western domination is still evident by the time this book is published, and what its global effects on cartography – and on how people view and represent the world in general – will be, remains to be seen.

what to show and how to show it. It is easy to forget this. It is easy to see mapmaking on a trajectory of ever increasing accuracy and objectivity. It is easy to see the arrival of technologies such as the Internet as heralding a new age of openness and equality, and it is easy to forget who pays for maps to be made and who is allowed and able to make them.

Maps have become a very common currency to use. People in western societies now consult maps so often that they often do not remember using them! These images have become commonplace, from bus maps to maps of offices, to maps in newspapers, in books and on the news. They are so common that we often fail to question their authority, fail to ask why they are drawn as they are. We routinely accept them as the 'truth'. The implications of this can be most severe when the maps are of places and people of which we otherwise know very little. We can criticize, with some knowledge, maps of our own times, neighbourhoods, towns or country; but maps of foreign lands often shape our perceptions with impunity. It is when we are presented with maps of far-off times or far-off places that we are at our most fallible and vulnerable, and it is when we are representing the people and places we know least about that we can do most damage.

At the end of the day, for all the power they contain, maps are just pieces of paper or merely ephemeral pixels (small rectangles of colour) on a screen. It is people who order, draw, purchase, use and learn from maps. And it is people who will improve them – not necessarily by making them more accurate but, for a start, by being more honest about how and why they are made and by teaching more carefully about how to read them. We can all be more open about why we make particular choices to map certain things, certain people and certain places. We can all think more carefully than we have done in the past about these things. We can also all look at other maps with a slightly more open and inquisitive mind and ask why the map shows what it shows – rather than try to merely understand how best the relationship between land and people can be painted onto paper.

Further reading

For an historical look and introduction to different ways of thinking about power and maps see A. Pred, 'Places as a Historically Contingent Process: Structuration and the Time-Geography of Becoming Places', *Annals of the Association of American Geographers*, 74 (2) (1984), 279–297, or Pred's *Place, Practise and Structure: Social and Spatial Transformation in Southern Sweden: 1750–1850* (Barnes and Noble Books, Totowa, New Jersey, 1986). The volumes in the *History of Cartography* series (edited by Harley and Woodward) cited at the end of Chapter 1 are the main references for the study of the history of cartography. Volume 1 (1987) includes G. M. Lewis' chapter, 'The Origins of Cartography', which discusses the propensities of different societies to create maps.

The 44th Monograph published under the *Cartographica* imprint (Volume 30, Number 1) in 1993 is aptly titled 'Introducing Cultural and Social Cartography'. This issue was edited by R. Rundstrom, and the *Cartographica* series is published by the University of Toronto Press. Articles from this issue of *Cartographica*, which have been cited above, include R. Rundstrom's 'Introduction' and 'The Role of Ethics, Mapping, and the Meaning of Place in Relations Between Indians and Whites in the United States'; D. Wood's 'Maps and Mapmaking'; J. Akerman's 'Blazing a Well Worn Path: Cartographic Commercialism, Highway Promotion and Automobile Tourism in the United States, 1880–1930'; B. Orlove's 'The Ethnology of Maps: The Cultural and Social Contexts of Cartographic Representation in Peru'; D. Cole's 'One Cartographic View of American Indian Land Areas'; and M. Edney's 'The Patronage of Science and the Creation of Imperial Space: The British Mapping of India, 1799–1843'.

Of Brian Harley's works, the most cited and perhaps the most accessible is 'Maps, Knowledge and Power' in D. Cosgrove and S. Daniels (eds), *The Iconography of Landscape* (Cambridge University Press, Cambridge, 1988). His 'Deconstructing the Map' and 'Cartography, Ethics and Social Theory' appear in volumes 26 (1989) and 27 (1990) of *Cartographica*, respectively, and he was also co-editor of the University of Chicago Press multi-volume *History of Cartography* series referenced above. For detailed obituaries of Brian Harley and further references, see M. Edney, 'J. B. Harley (1932–1991): Questioning Maps, Questioning Cartography, Questioning Cartographers', *Cartography and Geographic Information Systems*, 19 (3) (1992), 175–178; and W. Ravenhill, 1992, 'Obituary, John Brian Harley, 1932–1991', *Transactions of the Institute of British Geographers, New Series*, 17, 363–369.

Chapter 5

Mapping territory

Introduction

Chapter 1 revealed that map-making accelerated with exploration, with trade and with conquest, whilst Chapter 4 indicated that the mapping of indigenous peoples and the landscapes they inhabit was a primary aim of colonial powers. These activities, which require both the ability to control people, lands and routes and a supply of spatial information, will be considered further in this chapter.

Exploration needs, first, a coherent world framework; possibly a justification that the new lands encountered were *terra nullius* ('empty land') and hence eligible for occupation; and, finally, a means of recording and demonstrating successful colonization. Trade needs the geographical knowledge to exploit raw materials and acquire goods from distant societies; plus networks of agents, shippers and other contacts in far-off locations. For conquest, spatial information is as vital for waging strategic military campaigns and navigating over long distances to crush insurrections, as it is for the planning of local skirmishes and guerilla activity; insights into the nature of the land are essential to move armies and supplies; the remote delivery of deadly projectiles, from cannon-balls to cruise missiles, needs to take account of the spatial configuration of opposing forces and of the intervening landscape. In all these cases, the required information is perceived as ensuring control (often spatial information is kept secret to enhance such control) and it is for this reason that most of the map-making for exploration, trade and conquest has, in the past, been undertaken by government agencies.

A further reason for the preponderance of government activity in such map-making is its expense. Securing high-quality spatial data about exploration routes, trade possibilities and enemy positions requires considerable effort, often under trying circumstances and over considerable time periods. Clearly, only wealthy institutions could afford such activity.

Land ownership and mapping

It should be recognized, however, that there have also always been prosperous private patrons willing to commission map-makers. Early Roman maps often show the extent of a private estate, accurately delineated by the practical Roman land surveyors, *agrimensores*. Medieval landlords and squires were similarly responsible for initiating the large-scale mapping of substantial tracts of countryside. The resultant maps, although planimetrically well surveyed, were not working documents and seem to have been produced merely for self-aggrandisement. They also acted as symbols of authority: the addition of armorial bearings or personal decoration legitimized the right to possession of the land. The world seen from the atrium of a Roman villa, the battlements of a medieval castle, or the drawing room of a stately home was often limited to little more than these overviews of one person's landholdings.

Cadastral mapping

However, there has been a long-established category of large-scale accurate maps which *have* been used as practical tools since early Egyptian history. Cadastral maps are specifically created to help in the registration of ownership of land parcels. Such land registration is important because it pertains to the most vital resource which a society possesses. Half of the world's population makes its living through working on the land. One view is that all wealth comes, directly or indirectly, from land; and its efficient use is a primary aim of governmental planning. Hunting, fishing and grazing rights in land have often, in the past, been rather vague, both in terms of extent and of precise ownership, but with settled cultivation more carefully defined limits are necessary. The delimitation and documentation of property rights might encourage individuals to develop and improve the land. In addition, establishing such rights eases the transaction of land, stimulates investment and reduces litigation. From a government perspective, it is vital to know how much land exists within its area of authority, its use, its extent and subdivisions and its ownership. In many cases, the market value of the land is also recorded.

Such data can be stored as legal records in a public 'register', possibly in the form of tables, but the addition of a map, uniquely identifying each property, increases the utility and efficiency of such a register and turns it into a 'cadastre'. This link between the register and the map is not essential: some nations, notably Spain and in Latin America, have separate legal registers and cadastral maps; whilst other land registration systems, such as those used in the United States of America, although creating and depositing a plan of each individual property, may not use systematic and coordinated maps showing each property in relation to its neighbours. The cadastre does, however, improve the management of land registration systems and assure the unequivocal identification of properties.

Fiscal cadastre

A further primary role of government is to raise revenue. As land is a fixed and concrete asset it is a logical target for taxation policy. The cadastre can act as a fiscal register, allowing the governing authority to levy charges on land ownership and land transfer. To be equitable, such a system must have an accurate record of the extent of each individual's interest in land. As transactions involving land can be subject to some form of taxation, such as stamp duty, some means of ensuring that this is paid is necessary.

Legal cadastre

The cadastre can also be used as a legal instrument, designed to solve disputes about ownership of land. Such a role is important in areas where the value of land is especially high and it is in scarce supply. In such areas even a small discrepancy in the true position of a land parcel boundary could theoretically affect the price and usage of the land (although it is true to say that, in practice, the value of land is due more to its location rather than its precise extent). The legally binding nature of the cadastre is important, however: no transfer or subdivision of land can take place without it being recorded and certified. Information, in both the register and the accompanying map, therefore needs to be constantly updated. The usefulness of the cadastre as a current source of spatial data has not been overlooked by those implementing more general-purpose 'land information systems' which incorporate data additional to property records.

Land information systems

Cadastral mapping can often form the basis for widespread computer-based land information systems which are more multipurpose in application and which also primarily tend to be established in urban areas. Municipal mapping by local government has many functions – in engineering, development of power and communications utilities, management of public hygiene, implementation of planning legislation and maintenance of the stock of publicly owned properties: clearly the cadastre can form a core element in a computerized information system holding the records and spatial data needed to carry out these tasks (such systems are considered further in Chapter 7). When the map is drawn or the digital data are stored with reference to a national reference system or grid, cadastral maps can also be integrated with more widespread national topographic mapping (see pp. 90–93). This may prevent duplication of government mapping activity at large scales.

To meet the needs for accurate land registration, a number of systems have developed throughout history and in differing parts of the world. The link between register and map was first developed in France, and although such unification is not complete there, much of western and northern Europe has comprehensive and systematic coverage. Both Sweden and the Netherlands have significantly complete public records, mostly in computer-compatible form, describing a host of variables associated with property. In countries where fiscal cadastres predominate (e.g. Italy), notable extensions to the register in the form of building surveys have been undertaken and incorporated. Land reform in eastern Europe is an area of considerable contemporary activity: in some cases pre-war cadastres are being resurrected, but in others land is being looked at completely anew and legal redistribution of what were once state lands is likely to exercise lawyers and surveyors for many years to come! The privatization of land registration and cadastral mapping is farthest advanced in the USA (see Box 5.1), where the resultant non-systematic nature of records and maps has often led to calls for a more comprehensive and efficient system, primarily based on high-quality mapping. Differences in the methods of recording ownership (usually either by title or by deeds) and in the nature of the terrain have led to variation in practice in Commonwealth countries. In England and Wales, existing topographic mapping which records visible boundaries is sufficient to act as cadastral mapping – the property boundaries are long established and have been validated by owners erecting fences or planting hedges. In Australia, New Zealand and Canada, the Torrens system has been used which nominally requires maps for individual properties but in practice has been complemented by uniform, systematic mapping based on the topographic map grid. The use of existing mapping is common in many African countries where the jobs of topographic mapping and of cadastral mapping are undertaken by the same ministry.

Cadastral maps

In most cases, therefore, the cadastral systems are founded on accurate large-scale land surveys which, just as was the case annually in the Nile flood plain in ancient Egypt, allow for the boundaries of land to be re-established if they become destroyed. Some systems rely on noting accurate coordinated points along the boundaries of the land parcels, usually at the corners. This is a 'numerical cadastre' which ensures that a qualified surveyor (and in most authorities the surveyors who establish such coordinates and boundaries must be licensed) can accurately determine the extent of the land parcel – this is particularly useful when the land is being transferred to a new owner. Often the numerical cadastre is supplemented by actual marks (monuments) in the ground which indicate the extent of property boundaries. Numerical cadastres can be enhanced by accompanying maps which give a visual indication of the layout of the land parcels in an area. Usually such a map has inferior legal authority to the coordinate lists or the monumented marks and it is prepared for illustrative purposes only. It may be sparse in appearance with a monochrome representation of linear boundaries, sometimes with lengths and bearings of boundary lines, but with the minimum of topographic detail (Figure 5.1).

Certain systems, however, do rely on a map to embody the information defining the land parcel. Such 'graphical cadastres' are prevalent in long-settled areas where the boundaries of land parcels directly follow features on the ground which can be seen and mapped. Thus in England and Wales, as indicated, the boundaries of a land parcel are determined by its outline on the map and this invariably follows a feature, often long established, such as a fence, a kerb or a stream, which exists on the map. In some countries it is not even necessary to prepare a specialized map to depict such boundaries, as they are shown on existing topographic plans which can then be annotated as required by the registration authorities.

Applications

Cadastral maps and land registration systems are used

- for ensuring equitable and efficient taxation, maximizing the revenue to the state;
- for the administration of state works connected to the land (such as sewage disposal and power supply);

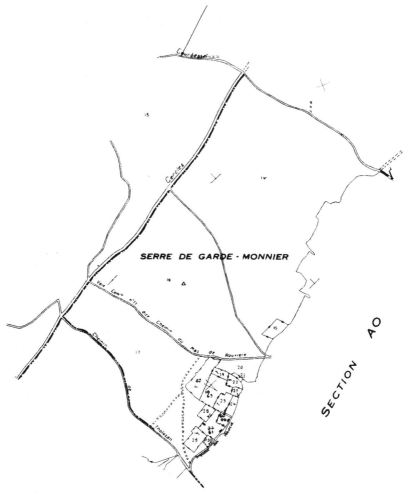

Figure 5.1   An example of a French cadastral map.

- to establish legal ownership, particularly in areas where orderly colonization of newly acquired lands (either by conquest or by land reclamation) was seen to be needed;
- to redistribute and enclose land and to implement agrarian reform;
- to establish and promote a land market by encouraging stability and investment;
- to allow the state to manage its own land resources for forestry, agriculture and mineral exploitation;
- to improve and extend map production.

All these activities are government-sponsored and the resultant cartography is politicized by the view of the authorities under whose auspices such important surveying and mapping is undertaken – the national interest is perceived as being met by the establishment of cadastral mapping systems.

Box 5.1   Contemporary mapping box – Settling the USA: the Public Land Survey System

The United States Public Land Survey System (PLSS) has been applied in 30 states (including Alaska) and was initiated in order to ensure the efficient sub-division of newly settled lands to the west of the Appalachians in the period after independence. The ordinance describing the methods for locating and disposing of such lands was passed in 1785. Townships were to be created, each 6 miles (480 chains) square, with 36 individual square mile (640 acres) subdivisions.

The initial impetus for the PLSS was the American colonists' hunger for land. Indeed the land issue was a major factor in forcing the War of Independence: a bitterly opposed Royal Proclamation of 1763 had reserved all territory from the Alleghenies to the Mississippi for exclusive use by the indigenous Indians. With subsequent independence, this land was settled, cleared, defended and, of course, mapped – hastening the decline of the traditional hunting grounds of the native tribes. Conflict between native and settler, focusing on land rights, has been continuous since that time. Withdrawing westwards in the face of colonial expansion, the native Americans were forced into ever more limited areas. The Allotment Act of 1887 recognized the possible extinction faced by such tribes. Unfortunately the response to this potential catastrophe was not tempered by an appreciation of the indigenous way of life: under the Act natives were to be granted title to land in 'Indian Territory', and to receive citizenship, provided that they would settle and engage in cultivation. As these were alien activities to the majority, such a scheme failed and 'Indian Territory', which was subject to considerable amounts of dispossession and neglect over the succeeding decades, was officially wiped from the map in 1906. This led to further land scrambles, and the decline in the area of native land was only halted by the Roosevelt administration in the 1930s. Acquisition of land to take account of expanding native populations has increased the area of native reservations gradually since that time, although the land is often of poor quality.

An enormous amount of land (1 800 000 000 acres; 7 284 000 km^2) needed to be recorded in the PLSS and the boundaries monumented. Although this was done under government jurisdiction (a federal surveyor-general was appointed from 1796), the vast majority of the survey work during the nineteenth century was done by private contractors, often unskilled and using poor-quality equipment, under pressure from settlers to work quickly and from hostile natives, frequently in difficult terrain and with lack of supervision. The low quality of work often led to enormous discrepancies in the field, so in 1910 a permanent body of surveyors was established.

All the required lines have been established at some time and the current interest of the government land surveyors is in the confirmation and occasional re-establishment of the survey marks. The subdivision of smaller sections is normally undertaken by local private surveyors. Since the PLSS was established,

(continued)

(continued)

two-thirds of the land parcels have been disposed of to individuals, commercial interests and state governments. The remaining land is still federally controlled; some (e.g. national parks and national forests) under specific authorities, and the remainder (272 000 000 acres) by the Bureau of Land Management (BLM, established in 1946). The land is used for recreation, forestry, wildlife conservation, grazing, wilderness area and cultural resource preservation. In addition, the BLM is responsible for rights of way, waste disposal, law enforcement, fire management and cadastral survey in these areas. Its rights to manage mineral exploitation concessions extend over nearly 600 000 000 acres in total. The federal government has always owned the title to native reservations, and the rectangular layout of townships and sections has had to take into account both these and natural features, notably bodies of water, as it divided the West up for the settlers. The subdivision of the land has been undertaken and managed in a systematic and efficient procedure which has led to the typical checkerboard patterns on the American landscape. Thus the dominant spatial patterns of the lives of millions of middle Americans have been determined largely by how this part of the world could be quickly surveyed 200 years ago.

Colonization and the subdivision of the earth

The concept of *terra nullius*, which was used to justify the westwards expansion of the United States of America, the settlement of Australia and the colonization of Africa, was only set aside by the International Court of Justice in 1973. Up to that time, and during the nineteenth century in particular, the world-view of the colonial powers was of ranges of primitive, unproductive terrain peopled, if at all, by savages or nomads who had no title to the land. Such areas were ripe for colonization, and the rivalry of nations such as the French, British, Germans, Italians and Spanish reached its height in the 'partitioning' of Africa which was ratified by the Berlin Conference of 1884–1885. The fascinating historical background to this episode of colonialism is detailed, involving diplomatic pacts, commercial interests, negotiated access to sea-ports and the relatively new concepts of free trade. In addition, a series of aims designed to ensure subordination of the local population was established which, although part-noble in spirit (suppression of the slave trade, protection of native peoples), led to a dismal humanitarian record of colonial repression.

The conference resolved certain disputes amongst the participating nations, notably between the Germans and the British, and with the international company set up by King Leopold II of Belgium to exploit the basin of the River Congo. Boundaries were to be established by raising flags, by negotiating treaties, by pressurizing local chieftains with shows of military might and by assuming protectorates. In virtually every case, the boundaries were determined in distant European capitals with reference to inaccurate mapping. Extending coastal zones of colonial influence inland was done using lines of latitude and longitude or other geometrical constructions. A series of now familiar straight lines appeared on the political map of the Africa as the ruler and compass were used to divide up territory on paper. Occasionally simple topographic features, such as watersheds or rivers were used, but in all cases verification of boundaries on the ground was problematic. Ignoring existing tribal frontiers, boundary determination often became a trial of imperial strength as national claims were exerted. Sometimes features were inaccurately located and one river, Rio del Rey, which was presumed to form the boundary between Nigeria and Cameroon was eventually found not to exist at all.

The rigour of geometric construction for dividing property and land, exemplified by the American survey system (see Box 5.1) and the African partitioning, has been applied for centuries at both this continental and at a local scale. The geometry of the graticule was used by Pope Alexander VI in creating the 'Tordesillas Line' in 1500, dividing the world into Spanish and Portuguese realms at approximately 45° west of Greenwich. In ancient times many Roman cities, and their Chinese counterparts, were planned with city streets laid out on a grid system. The 'purity' of the grid, often overlaid by the architect on topography as yet unseen, has been applied to planned cities such as Philadelphia and Melbourne, Kyoto and Chandigarh, since that time.

Contemporary mapping is still of crucial importance to the subdivision of land for political gain. For instance, in the United States of America, land that was originally subdivided under the concept of *terra nullius,* is now regularly reapportioned every ten years in a ritual that follows the release of data from the decennial census of population. Mark Monmonier (1995) discusses the connections between mapping and politics in detail in his book *Drawing the Line* in which he shows that when population shifts and the constitution require a redrawing of legislative and congressional boundaries, reapportionment gives the party in power enormous latitude to consolidate its gains, help incumbents, and satisfy the Voting Rights Act. The technique first attributed to Governor Elbridge Gerry remains as influential today as when the term 'gerrymander' was invented following his manipulation of political boundaries within Massachusetts in 1812. Monmonier shows that 'boundaries redrawn after the 1990 census proved particularly contentious in North Carolina and other states where efforts to satisfy minority-rights watchdogs at the Department of Justice led to contorted Congressional districts . . . and charges of "racial gerrymandering".' Following the results of the 1994 national elections, it became apparent that the efforts to empower African-American and Hispanic voters by (legally) designing constituencies in which they would form a majority actually bolstered the position of the more conservative party in America by increasing the number of 'wasted' ethnic minority votes. Maps are useful both to help in the gerrymandering process and in helping to expose it. A sure sign of gerrymandering is the 'salamander' constituency, a political zone on the map the boundary of which is so contorted that it can resemble an attempt to draw a picture of an animal out of minor areas. In one celebrated case the zone had so many appendages added and subtracted that it appeared that the boundary commission had contrived to drawn a salamander on the political map of their country.

The impact of the military on mapping activity

Just as governments are willing to invest heavily in a comprehensive land registration system, usually with the potential aim of generating tax revenue later, so it often appears that there is little restriction on expenditure on map-making by the military arm of government. The return on this investment may be perceived as losses in battle minimized, rebellions suppressed, colonies won and control reasserted. Maps have a pivotal role in war although many peace-time maps originated in mapping for military campaigns. Contemporary topographic mapping, in particular has obvious roots in the military activity of the nineteenth and twentieth centuries.

Maps in the war room: the impact of contemporary military technology

Despite the ubiquity of the global news crew, the primary representation of the world at war is still through the medium of the map: from subtly courting public opinion using a map of tiny Kuwait being engulfed by the military symbols of the massed hordes of Iraq, to the crudity of a cruise missile following detailed contours to its grid-referenced target; from the excitement of the TV pundit poring over the battlefield plans, to the hourly briefing provided to the com-mander-in-chief.

In addition, it is still true to say that most contemporary advances in map-making technology are triggered by military considerations. Military activities require sophisticated access to topographic information in a variety of forms. Unfortunately, the freneticism of battlefield activity is unlikely to lead to coher-ent and sober application of such information. Imposing order on the chaos of a military engagement is often unsuccessful. Clearly, from a military point of view, maps and spatial information consulted must be precise, concise and capable of being acted upon in the heat of battle. More often it is activities mar-ginal to combat which rely on map products: helping escapees; training recruits; positioning oneself preparatory to conflict; establishing lines of communication and supply; controlling civil unrest; and creating propaganda for the populace back home.

The legacy of military cartography

Topographic mapping depicts the characteristics of the visible landscape around us. Natural features, such as hydrology, land cover and vegetation type, are major elements of topographic mapping: it is clear that such elements have an impact on the engagement of battle in a location. Similarly, human features (often called 'cultural detail' in cartography) are important items to include on any map, as they help to detail, for a soldier, the nature of an area, routes to use in navigat-ing through or around it, and the size and distribution of the local population. Certain elements such as contours, boundaries and geographical names are not necessarily directly visible in the landscape itself, but they are essential

Box 5.2   Contemporary mapping box – Mapping the Gulf War

When preparing for active service behind enemy lines during the Gulf War in 1991, the SAS soldier Andy McNab was appalled by the topographic information available to him initially (McNab, 1993). An example of the type of paper map data supplied to his unit is reproduced below. Clearly, any additional information about the terrain, preferably at a larger scale, would have helped the cause of soldiers engaged in sabotage operations or trapped within Iraq.

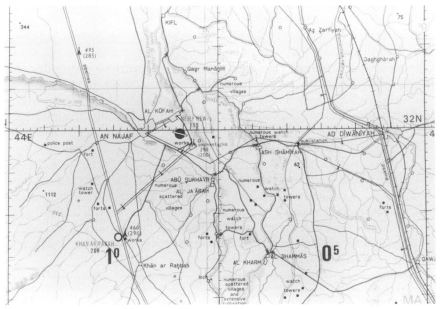

Extract from Tactical Pilotage Chart TPC H6AG 1:500 000.

By the end of the war, the Allied forces (although not the unfortunately captured Sergeant McNab) were using terrain information of extremely high accuracy and completeness to help overcome the Iraqi Army. In this arena of war, as in other contemporary operations, such as Bosnia, much of the combat was undertaken not on the ground, but in the air and from the sea. A bewildering array of spatial data, most in digital format, are available to the modern-day commander and the foot-soldier, and the techniques of map-making and map use are equally dependent on the availability of computer power. Much of the modern technology used to promote war relies on the type of terrain data exemplified by the list in Box 5.3.

Such data can be combined with

- the availability of high-precision positioning devices using GPS (see Box 3.1) – more than 14 000 GPS receivers were used by Allied forces during the Gulf War;

(continued)

(continued)

- almost real-time satellite imagery of high resolution, allowing for rapid-response map production of photo-maps (more efficient than conventional topographic line maps for desert and many other terrains); and
- on-site image processing and GIS facilities.

The result is a flexible battlefield information system, with terrain information at its core, but with applications in all aspects of modern warfare – planning, logistics, command, control, support and combat.

More specific elements of such an information strategy are essential to ensure the efficiency of a range of modern weaponry. The Pershing cruise missile relies, for its accuracy, on holding a digital terrain model of its entire intended flight path such that it can 'hug' the ground and penetrate enemy lines below the radar defences. On-board navigational aids are necessary to ensure that the missile is travelling correctly towards its intended location. In order to pinpoint its objective exactly, it requires a simulated radar image of the scene which is then pattern-matched with the radar picture sensed in reality to ensure precision targeting. The required developments in spatial technologies to ensure such operations are clearly extraordinary.

This matter-of-fact description of the operation of a cruise missile hides the fact that real people are involved in its development, construction and use. The de-humanizing influence of war and all the activities required for its prosecution are emphasized as much by a reliance on readily available seamless digital terrain data for the whole theatre of war, as by any of the travesties of the English language employed by the military spokesmen relating 'body counts', 'collateral damage' and 'friendly (sic) fire incidents'.

denotations of its shape and configuration. The content of topographic maps has, in fact, changed little since the establishment of national governmental agencies charged with producing them primarily for military activity in the eighteenth century. Standardization of content is such that topographic maps from widely differing landscapes, produced by different national agencies, employ notably similar symbolization. Standardization over time means that the legend of a 1990s 1:50 000 topographic map would have been understandable and usable (apart from the detailed road and railway classification) by a 1790s cavalry officer (Figure 5.2).

In addition to its government sponsorship and its conservatism, a further notable characteristic of most topographic mapping is its uniformity of style, content and appearance. Uniform 'map series' consist of a potentially large number of map sheets, each of similar size and layout, representing a substantial portion of land, quite often the entire nation. Such a map series ensures that the military commander has detailed large-scale coverage of specific areas, yet also has a host of adjoining map sheets onto which his campaigns or exercises can be moved if necessary and from which an overview of large tracts of landscape can be obtained.

The world-view of a military officer is governed by the landscape he sees before him – and the mapping has been done in the same way: direct observation of the landscape by survey parties, by using aerial photography and by expensive and accurate instrumentation in the office. Topographic mapping to a uniform standard, both to create new maps and to maintain existing maps by revision (clearly an important task), is expensive.

The word 'he' in the previous paragraph is appropriate because a further point related to the military legacy of cartography is that as a result of the prevalence of such activity in the past, the map-making profession is still very much male-dominated. Most of the senior positions in surveying and map production organizations and the majority of written work on the subject (including many major textbooks written by ex-military officers) are controlled by men. This is despite the widespread employment of women in more menial cartographic activities, such as drafting and computer operation. The position of women in cartography is improving slowly, although at differing rates throughout the world. The International Cartographic Association (ICA), a forum for national cartographic societies throughout the world, has a goal of promoting equality of opportunity in all organizational units and at all levels of responsibility within its member countries. An important step has been the initiation of an ICA Working Group on Gender and Cartography which has published a Directory of Women in Cartography, Surveying and GIS, containing data on 400 women from over 40 member countries. This has been used to increase participation of women in cartographic conferences and to assess circumstances and opinions of women cartographers employed in map-making activities. Whether such developments will radically change the nature of mapping and map-making is uncertain, although it seems highly likely that the prevalent ethos in cartography will remain masculine.

Contemporary government mapping

The military nature of topographic map production and potential use means that, even today, a significant number of national mapping agencies produce maps that are not allowed to be distributed to the civilian population. Large-scale maps, which accurately locate strategic resources such as dams, power lines, urban centres, port facilities and military airfields are often regarded as sensitive documents which should not be used by any other than the national armed forces. Some agencies are still completely staffed by military personnel.

However, a new agenda has overtaken many such national mapping agencies and they are being forced to respond to the demands of government to be more cost-accountable. In many cases the armed forces have long since ceased to be the major customer for the agencies' products and a civilian strategy is being followed. The value of the nationwide topographic information controlled by the government map-making authorities should, it is suggested in these expenditure-aware days, be maximized. This requires the agencies to become much more market-oriented and seek to actively promote the use of their topographic database, whether in paper map or newly available digital form. In addition, certain governments have established policies which enshrine the freedom of the

Figure 5.2A Example of Ordnance Survey topographic mapping of 1801 (covering part of Kent, England).

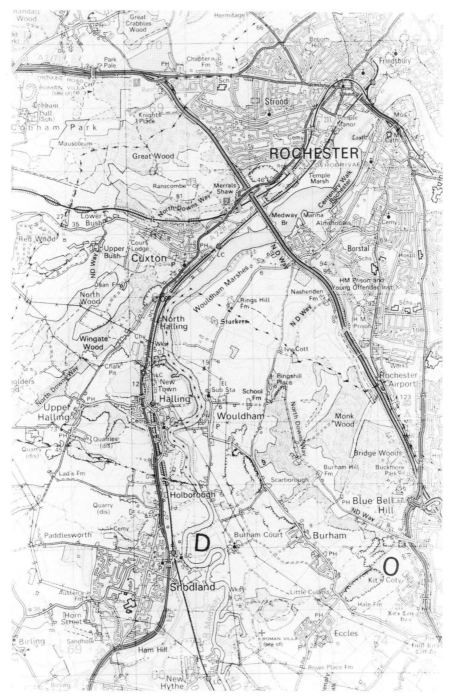

Figure 5.2B Example of Ordnance Survey topographic mapping of 1990 (covering part of Kent, England). © Crown copyright.

Box 5.3   Contemporary mapping box – The NIMA inventory of digital spatial data

Access to high-resolution, high-accuracy, current and complete digital data sets over wide areas is a prerequisite for waging war, particularly in an era when regional conflicts seem likely to occur anywhere in the world. The National Imagery and Mapping Agency of the United States Department of Defense is 'meeting the challenge' of supplying such data by moving towards a 'global geospatial information and services' strategy, committed to shifting from a focus on paper map products to a concentration on digital spatial data. Operational requirements dictate that a significant range of terrain information, plus some data on infrastructure, is available for the armed forces. Many of these data are compiled from existing paper mapping, but significant amounts are obtained from high-resolution satellite sensors (including radar), aerial photography and subsequent image processing.

Among the vast range of data sets available are:

- ARC (Arc Raster Chart) Digitized Raster Graphics (ADRG): raster (grid-cell) digitized paper mapping;
- Controlled Image Base (CIB): fully geocoded (i.e. transformed to a reference projection and grid) panchromatic satellite imagery, corrected for relief distortion;
- Controlled Multispectral Image Base (CMIB): multispectral satellite imagery complementing CIB;
- Digital Bathymetric Data Base (DBDB): gridded ocean depth data at 5 minute (latitude/longitude) interval (equivalent to 10 km);
- Digital Chart of the World (DCW): 1700 Megabyte vector (line mapping) topographic data set obtained from 1:1 million scale maps;
- Digital Feature Analysis Data Level 2 (DFAD2) and Level 3–C (DFAD3-C): selected natural and human-made planimetric point, line and area features held in a vector database obtained from air photography and large-scale maps, at variable resolutions; it is combined with DTED2 to give simulated radar pictures for pilot training and terrain matching;
- Digital Nautical Chart (DNC): vector data from hydrographic charts – the nautical equivalent of the DCW;
- Digital Terrain Elevation Data Level 2 (DTED2): 1 second (latitude/longitude) interval (equivalent to 30 m) digital elevation model;
- Tactical Terrain Data (TTD): the standard land combat data set comprising complete terrain information for specific battlefield situations and armed forces operations; this is a layer-based set of topographic data;
- Vector Smart Map Level 2 (Vmap2): high-resolution vector data sets separated into 10 thematic layers – an enhancement of DCW, including some DNC data also;
- Vector Smart Map Urban (UVMap): high-resolution vector data sets separated into 10 thematic layers with city graphic content – an enhancement of DCW, including some DNC data also;
- World Vector Shoreline–Vector Product Format (WVS-VPF): standard format vector coastline data set with offshore boundaries.

Box 5.4 Personality box – William Roy and the Ordnance Survey

Like many renowned British mapping engineers, William Roy was born in Scotland – at Miltonhead, Lanarkshire, on 4th May 1726. His father was a land agent and William probably accompanied him on survey work undertaken for the lairds of the Milton estate. After leaving school at the age of 14, Roy moved to Edinburgh, gaining experience as a civilian draftsman with the Board of Ordnance at Edinburgh Castle. His immediate superior, Lt. Col. Watson, Deputy Quartermaster of the British Army in Scotland, having witnessed the inefficiency of the maps used in finally putting down the Jacobite rebellion in the Highlands of Scotland in 1746, suggested that accurate surveying and mapping be carried out, complementing the programme of new road building in the area. Newly promoted to Assistant Quartermaster, Roy was charged with the project, leading six survey teams. From 1752, the map-making was extended to cover the whole of Scotland, but three years later the start of war with France curtailed the work and Roy was sent to the south coast of England, also seeing active duty on the continent.

Roy's view was that an accurately surveyed control scheme, along with adequate maps, were essential for the orderly promotion of government. He first proposed a national survey, with a similar remit to the Cassini-led organization in France (see p. 20), in 1763. Despite re-submitting a less ambitious plan in 1766, the project was thwarted by its perceived high cost. Roy continued mapping fortifications and undertaking engineering surveys for the remainder of his army career, although he still held a view on the national utility of topographic mapping. In 1783, he independently initiated a triangulation scheme in London, designed to fix the positions of the major church steeples and public buildings. Such activity was also intended to remind both the public and those in government of the advantages of a nationwide triangulation and mapping scheme.

An invitation from Cassini III to help determine the relative positions of the Paris and Greenwich observatories was delivered to Roy in 1784 and the work was initiated with great enthusiasm. Roy persuaded a number of prominent patrons, many fellow members of an informal Royal Society dining club, of the importance of such a project, which required the establishment of an accurate triangulation scheme over a considerable stretch of country. The measurement of the base station on Hounslow Heath is regarded as the start of the triangulation of Great Britain. Despite bad weather and Roy's deteriorating health, the scheme was completed in 1787 and the link across the English Channel to the French control network was established (see figure). The high accuracy of the two different schemes (the British and French estimates, obtained from the triangulation schemes started at Hounslow and Paris respectively, of the length of the French check line laid along the coast near Calais varied by only 1.25 feet; 0.38 m) confirmed the utility of a triangulated control network of accurately established points. In a posthumous paper (he died in 1790), Roy proposed the

(continued)

(continued)

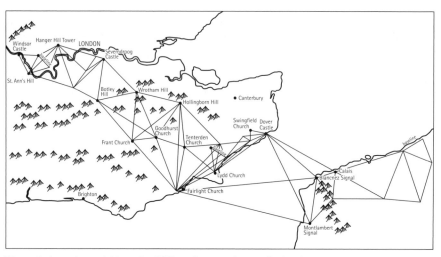

Triangulation scheme laid out by William Roy, south-east England, 1784–1787.

extension of the scheme across the whole country. He did not live to see the creation of the Trigonometrical Survey (the original name for the Ordnance Survey) in 1791, but his legacy is a surveying and mapping institution of high repute whose spatial data products owe much to his foresight and determination.

citizen to have access to virtually all information collected by government. Such information obviously includes topographic map products and here again the old notions of secrecy are being rapidly dismantled.

In some cases policies of 'openness' and freedom of information are being offset by the increasing monetary cost of spatial data to the average citizen. The market-orientation of some mapping agencies, notably the national agencies in Great Britain, New Zealand and France, leads them to charge considerably higher prices for the supply of contemporary spatial data than were charged in the days before cost-accountability. These are prices which are often beyond the reach of the average citizen, although they are realistic because they are the market rates which can be paid by large commercial companies. Such pricing of spatial topographic data, particularly data in digital form, has ensured that it is the national mapping agencies mentioned which are able to invest in new equipment, supply the information which the consumer wishes, and remain in the forefront of contemporary developments in mapping and cartographic data handling. Certain other governments, which have a less aggressive commercialization agenda for national topographic information, tend to supply spatial data which may be out-of-date, incomplete in terms of content or non-uniform in area of coverage.

In either case, however, it is noteworthy and ironic that the openness of government in many societies when supplying topographic information means

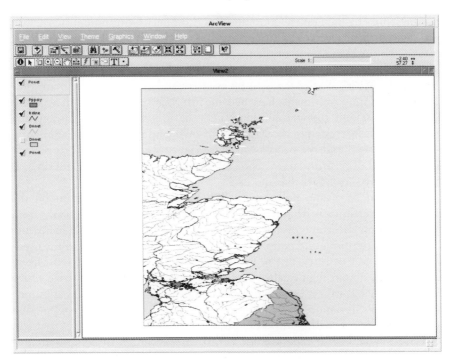

Figure 5.3   An extract from the Digital Chart of the World viewed in the ArcView GIS software.

that it is the topographic map (originally designed for and used by soldiers) which is the most readily available to the general public; whilst maps which affect the population's everyday lives (planning maps, geodemographic products and transport network maps) are inaccessible or complex. Youth groups such as schoolchildren are introduced to maps through topographic map reading exercises; the topographic map is the cheapest, most readily available tool for interpreting the landscape in school exercises; the outdoor leisure enthusiasts, encouraged by advertising campaigns to take advantage of the 'great outdoors' are informed that a topographic map is the most important piece of equipment they can carry with them; the geography student who wishes to produce a regional map for project work may have recourse to the data provided as part of the Digital Chart of the World (DCW), a digital data set sourced from military topographic map data and accessible from GIS packages (see Chapter 7) such as ArcView (Figure 5.3). The view of their immediate locality which most citizens obtain is through the window of a product which originated to serve the needs of a small group of late eighteenth-century military decision makers.

Summary

The role of government in mapping has been crucial for its development, from the initiation of land recording surveys in Ancient Egypt; the patronage of the

Cassini family by the French Court and Revolutionary Government (see p. 20); the establishment of colonial survey organizations such as the Survey of India (see p. 73); the burgeoning of the mapping programme of the CIA; to the use of spy satellites to map troop movements in wartime. The representation of the earth by legislative bodies tends to be assertive and complete. The maps produced by and for governments have become information archives with a nationalist, militarist and topographic bias. They have confirmed the notion that government is interested in the nature of its own land resources, the configuration of its terrain and the people who have rights over land within its boundaries. Control over information about the land (and indeed the sea and skies) is necessary for control over the land itself.

We should be aware, however, that the influence of government on the mapping of territory is decreasing. Civilian sources of data, particularly from remote sensing satellites (covered in more detail in Chapter 6), are of increasing accuracies, resolutions and consequent use. If we cannot get access to the topographic map of a country, due to its restrictive distribution, we are most likely to be able to obtain high-quality satellite imagery of the area relatively easily. Such imagery has uses when waging war (see Box 5.2), but is also as appropriate for tourists in remote areas, fishermen searching for fish-shoals, population census enumerators and pollution monitors.

Unfortunately, such data are not cheap, and although in certain circumstances altruism dictates that the data be made available at low cost to all who have need of them (this pertains, in particular, to the United States Landsat programme), there is a danger that satellite imagery can be misused in the commercial sphere. Mineral exploration companies armed with satellite data and sophisticated interpretation facilities are, for example, quite capable of deceiving naive national governments when negotiating exploitation rights and accounting for their actions and proposals.

There is currently a shift in the balance of information holding and perhaps, therefore, in the future control of territory. As the activities of the state (including mapping) are being 'rolled back' in most post-modern societies and replaced by private enterprise, 'commercialization' of cartography has led to a revival of the private sector in mapping which harks back to the 'Golden Age' of seventeenth-century Dutch map-making. In addition, the 'civilianization' of the mapping of territory is proceeding apace: Chapter 8 will discuss the impact that particular local and citizen initiatives are having in the mapping process. In the meantime national mapping agencies struggle to maintain their identity in an era where, although still vitally important, spatial information is wanted more quickly, more cheaply and in radically different forms than previously.

Further reading

The story of a national mapping programme and a national mapping agency is exemplified in considerable detail in N. L. Nicholson and L. M. Sebert, *The Maps of Canada* (Dawson, Folkestone, 1981). Map-making and map use activity connected to wartime activity is illustrated in a series of vignettes which make up Chapter 4 of P. Barber and C. Board, *Tales from the Map Room* (BBC Books, London, 1993). Technical details of the United States Public Lands Survey System can be found in most American land surveying text-books, such as P. Wolf and R. C. Brinker, *Elementary Surveying*, 9th edition (HarperCollins, Hinsdale, Illinois, 1993). The role of mapping in land registration throughout the world is covered in an historical survey by R. Kain and E. Baigert, *The Cadastral Map in the Service of the State: A History of Property Mapping* (University of Chicago Press, Chicago, 1992), whilst the use of land registration systems for management purposes world-wide is described by G. Larsson, *Land Registration and Cadastral Systems* (Longman, Harlow, Essex, 1991). The specific example of nineteenth-century African colonial boundary determination is briefly described in Chapters 17 and 20 of S. Forster, W. J. Mommsen and R. Robinson (eds), *Bismarck, Europe and Africa* (Oxford University Press, Oxford, 1988). A useful history of the Ordnance Survey, which highlights Roy's role in its establishment is T. Owen and E. Pilbeam, *Ordnance Survey: Map Makers to Britain Since 1791* (HMSO, London, 1992).

Chapter 6

New scales, new viewpoints

Introduction

In recent decades, 'mapping' has taken on a new meaning in science, far wider than the representation of the surface of the earth and objects thereon. Human genes are now mapped, with the guiding metaphor for the international Human Genome Project being to create an 'atlas' of what makes us human. Computer graphics have allowed mathematical functions to be mapped visually as well as functionally, cumulating in the drawing of glorious landscapes which look real but are simply the graphic realization of the solutions to a one-line equation. Whole planets and even large parts of the universe are currently being mapped in ways that are a million miles away from the ancient sky mapping of astronomy. From the smallest objects to the largest, from the most stable to the most abstract, scientists from many disciplines (which traditionally did not draw any pictures) are now drawing maps. In order to interpret these objects as maps and fit these developments into the story of mapping, an appreciation of what is different about these new ways of representing the world is required. That is the purpose of this chapter.

The context for these developments is the growth of what has come to be called 'scientific visualization'. Visualization is the process of learning through the creation and observation of abstract images, providing a method for seeing the unseen. 'Scientific visualization' was announced to the academic community as well as to the rest of the world in 1987 by a report of leading American scientists (McCormick *et al.*, 1987a, b) which received widespread publicity from the popular as well as the academic press. The aim of visualization is to take the graphic beyond the realm of illustration. It differs from illustration in that the purpose of visualization is to *discover* the unknown rather than to *show* what is already known. Its origins in science can be traced back far beyond the current computer graphics-inspired renaissance (the word was used by Descartes, for instance). New technologies have been developed over the centuries to help draw graphs and illustrations and they have come into fashion, only to fall out again once such technologies and the pretty pictures they had created became no longer novel (Beniger and Robyn, 1978). Some scepticism may well, therefore, be warranted as to the novelty of recent developments. There were overt political motives behind the original report on scientific visualization; for instance, to maintain the funding of expensive super-computing facilities after they were no longer needed to simulate atomic explosions and to aid the

American computer industry which thought at that time that it had already lost the market for the production of hardware to Japan. Despite these reservations, one only has to consider some of the images now being created in the field of scientific visualization to gain the impression that we are able to draw and view images of the world that were literally inconceivable a decade ago.

In terms of the effect on cartography and the creation of new representations of the surface of the world, it is worth remembering that, although there are many aspects of the new mapping which are revolutionary, there have always been objects that were on the boundaries of being considered maps. Some of these, in particular perspective views of the world, are also considered here. Although the world has changed, in that many more maps of many different kinds and of many different objects are now being produced, cartography, and the sciences in general, have also changed to reflect more pluralistic fields of enquiry and to follow practices that were previously on the boundary of what was once permissible to study or report. Thus, a physicist can now draw and publish maps of, say, particles being analysed, whereas a few decades ago it would have been considered amateurish to publish such pictures in a scientific article. Similarly, a cartographic journal may now include an article on the cartography of the atom, a subject that would have been deemed firmly outside its scope in the past. Thus, how our representations of the world change over time depends as much on how the norms of what is 'allowed' alter (depending on the imaginations of the reader) as it does on the imaginations of the people who are to draw the new pictures. Given the terms defined in the Introduction, many of the images discussed below are map-like objects rather than maps. Similarities can therefore be made between the drawings of ancient civilizations and the current products of high technology. They both stretch the boundaries of what can be considered as maps.

The new world-view: an alternative icon

A whirlwind tour of the world captured by modern mapping extends from the atomic and microscopic to the cosmic. Planetary geologists have mapped the hills and dales of Venus by radar, Mars by magnetometer, Jupiter with photo-polarimeters, the Moon in person. Paleoecologists have mapped the location of lakes that dotted the Sahara until disappearing four thousand years ago, and climate modellers are mapping the climate as it will appear one hundred years hence. From 590 miles up in space, satellites can determine the average income of a neighbourhood, follow wandering icebergs, track the wandering albatross; from instruments resting on the surface of the earth, physicists can see into the heart of the planet, into the heart of the atom, into the heart of the Big Bang. (Hall, 1992, p. 7)

Such is the stuff of a science journalist's dreams. To the sceptical student who knows a little about these subjects, much of the above may appear fanciful, but there is truth in all these assertions. Venus may not have hills and dales as we know them, but there are cartographers of (and now several PhD theses in cartography and geodesy on) extra-planetary mapping. A satellite may not be able to determine average income, but by measuring the density of buildings in a suburb, and knowing how far that suburb is from the urban centre, a close correlation with average income can be made. As the revolution in computer visualization unfolds, the map becomes an increasingly useful (old) metaphor for the new science. In one sense, all that has changed is that whereas in the past it took a great deal of human effort to produce just one conventional map of a town, there is now abundant surplus human effort looking for new things to map. In other senses, much more has changed than this.

If all these new pictures are to be accepted as maps then a new definition of map-making is required. Stephen Hall, the author of the above quotation, proposes a simple one: 'map-making is the process of winning data points from nature' (Hall, 1992, p. 8). After 'winning', he might perhaps have added 'and representing', but that is to quibble. In universities across the world academics have come to accept that there is much more to map than merely the surface of the earth, and that the methods which have traditionally been used to map the surface of the earth can be very useful for mapping other things. How did this transfer of skills take place? One route we can follow, to begin to understand this recent history, is towards the conjunction of cartography with satellite remote sensing. In remote sensing, computer-generated images are used to create pictures of the earth – pictures which, at first, did not look all that familiar. Furthermore, the techniques that were used in remote sensing could be used to study any other information which could be ascribed and digitally encoded to a grid of pixels.

Remote sensing and data

Several examples of the utility of remote sensing in cartography were given in Chapter 3. In essence, when first introduced, remote sensing techniques were seen as having three advantages over more traditional methods of cartography (Rabenhorst and McDermott, 1989). First, they simplified the identification of objects that were to be placed on maps. Secondly, they allowed maps to be revised more easily and for the urgency of more accurate revision to be estimated. Finally, the digitally encoded, remotely sensed images could be used directly in the computer-assisted map production systems being developed in the 1970s and 1980s. However, the potential uses of remote sensing extend far beyond those of traditional aerial photography which has long been used in topographic map-making. Levels of heat rather than light can be detected and used to monitor environmental change or, for instance, the proportion of homes in an area which appear to have roof insulation. To understand what remote sensing is capable (and not capable) of, a brief understanding of the technology is necessary.

Remote sensing systems can sense visible, ultraviolet and infra-red light, but only certain wavelengths of this spectrum can pass through the atmosphere relatively easily and, of course, the atmosphere varies in its transparency to such wavelengths (the most simple example being that clouds interrupt the view in the visible part of the spectrum). Different satellite sensors produce different images and the range of these is expanding with each new remote

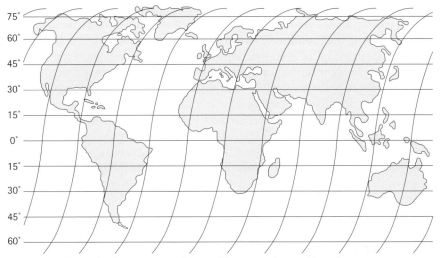

Figure 6.1 The path of a remote sensing satellite across the earth's surface.

sensing mission. Many sensors capture their digital images at differing wavelengths simultaneously, each image covering an area of the earth's surface (its 'footprint'). For the type of image most suitable for use by map-makers, the footprint is up to 200 km wide and tracks across the earth's surface as the satellite orbits (Figure 6.1). The spatial resolution (i.e. the size of the pixel that is sampled) varies from a few tens of metres to a few metres square, although some resolutions are considerably more coarse than this depending on the wavelength sampled and the height of the satellite above the earth.

The costs of obtaining these data are not low, despite commercial competition in the skies (the American Landsat system has been supplemented by the French SPOT satellite, and numerous specialist sensors, particularly sensing radar images). If an image of a particular place is required to be made, at a time when minimal cloud cover exists, this is possible using acquisition programming techniques, but the cost is even greater (see Further reading, in particular Rabenhorst and McDermott, 1989, p. 28).

The technical aspects of handling remote sensing imagery can become quite complicated. Operations such as converting the complex geometrical properties of the image to a traditional map projection and the enhancement of an image to improve its interpretability using particular computer algorithms are not trivial, although they are often embedded in specialist processing software. Learning to interpret pattern, texture and shape from remotely sensed images is not as easy as it first appears either. Thus, not only is there a high cost to the collection and dissemination of data from imaging satellites, but the analysis of that data to create a useful product also comes at a price that is often underestimated. Knowing this, it may not surprise the reader to learn that the vast majority of remote sensing is conducted *in* the richer countries of the world, although this is often *of* the lands of some of the poorest people. For instance, remote sensing allows 'resources' to be identified and measured at the same time as greatly aiding the monitoring of environmental degradation. It is now possible to prospect for oil or even gold with the initial explorations being done by satellite. Government can have millions of private maps – maps which could not be drawn before – made relatively cheaply (for them). Such maps may well be of other countries. But it is, perhaps, in its military uses that remote sensing has had its most dramatic impact. Recent well-publicized examples range from the pictures projected on the world's television screens showing the supposed mobilization of aggressive forces on a country's border, to the private, classified images which are used to make and justify key decisions over people's lives and the future of thousands of communities (see Box 6.1). The use of remotely sensed data can imply even greater costs again than those involved in its capture or analysis.

Scale and accuracy beyond belief

> Until the early twentieth century trade was usually the motive although competition frequently took a military form. In our own times it has been the needs of defence and warfare which have powered and funded the development of navigational science. (Williams, 1994, p. 1)

Williams begins his recent book with a powerful argument about the interrelationship between technical change and ways of representing and understanding the world – particularly about where we are, on or off it. Navigation grew dramatically in scientific importance after the 'discovery' of the Americas, when it became increasingly important for those piloting ships in the open ocean to know where they were. As is described in detail in Chapter 3, the solutions that were developed had a dramatic impact on the ways in which cartographers drew maps for centuries later.

However, once the positioning problem had been solved to an acceptable level of precision, developments in navigation decreased in importance and scientists tended to concentrate on other things. Then, at the turn of the twentieth century everything changed as the first aeroplane was flown. Suddenly it became dramatically important to know, and to be able to show, very precisely where you were: how else would you know where to land?

Just as the development of satellite imagery first required radio to transmit the captured digital pixel values (and in time began to be used to explore what could be sensed using wavelengths near radio), it was radio that revolutionized navigation this century. A radio signal provides a beacon that can be followed and steered towards – like a lighthouse, but with one obvious advantage: you do not have to be able to see it to steer by it. Thus navigation from afar, in the dark or in fog is possible when following radio signals and this is what was essential for navigation in flight. One of the simplest early systems was built by Telefunken using 32 transmitters which, like a set of lighthouses, transmitted signals in different directions at different times (Williams, 1994, p. 187). By

Box 6.1 Contemporary mapping box – Pictures from space: the military and mapping

The history, development and much of the current use of remote sensing has been conducted, largely in secret, in the military domain. Remote sensing has superseded paper-based cartography for many military uses. Remotely sensed images now contain the secrets which others must not see and it is increasingly remotely sensed images that are used by political and military leaders in making decisions. Their paper-based maps tend no longer to contain information that is not generally available.

Rabenhorst and McDermott (1989) argue that remote sensing exposes ground truths which can be used to avoid conflict. They cite, as an example, the aerial imagery collected by U-2 spy planes above Cuba in 1962. These pictures showed missile sites being built, and resulted in the ultimatums and blockade which collectively formed the Cuban missile crisis. Whether this crisis was an example of 'avoiding conflict' is debatable. It is easier to argue that remotely sensed images have played an increasingly important role in war over the last few decades.

Contrast the following two quotations, the first from a remote sensing textbook, written by an American colonel who was part of a photo-interpretation team in south-east Asia in the 1960s, and the second being the start of a journalist's account of what happened on the ground:

> Frequently a crucial military operation is held up, pending the results obtained from image analysis. That imagery, be it hand-held or from an exotic remote sensor, is precious, and so is time. That makes the PI (photo-interpreter) like an umpire or referee. A look, and then the call must be made. Has the target been destroyed, or must we bomb again tomorrow? . . . During the Southeast Asian conflict, the very best PIs could also rattle off the six-place geographic coordinates and target number for most of the hot objectives. (Colonel R. M. Stanley II, quoted in Rabenhorst and McDermott, 1989, pp. 105–106)

> On February 9, 1969, less than a month after the inauguration of Richard Nixon, General Creighton Abrams, commander of United States forces in South Vietnam, cabled General Earle G. Weeler, Chairman of the Joint Chiefs of Staff, to inform him that 'recent information, developed from photo reconnaissance and a rallier gives us hard intelligence on COSVN [Central Office for South Vietnam] HQ facilities in Base Area 353 [in Cambodia]'. (Shawcross, 1986, p. 19)

Following on from the second quotation above, as a result of the photo reconnaissance and on the American President's orders, in the early hours of 18th March 1969, sixty B-52 plane loads of bombs were dropped into Cambodia, a country which America claimed to regard as neutral. Before the bombing, code-named 'Operation Breakfast', 1640 Cambodians lived in Base Area 353, of whom the Americans estimated at least 1000 to be peasants. Operation Breakfast was followed by bombings labelled as Operations 'Lunch', 'Snack', 'Dinner', 'Dessert', then 'Supper'. The American people did not find out what their forces were doing on their behalf until 1973.

(continued)

(continued)

The photo-interpreter of the 1960s and the remote-sensing analyst of the 1990s may have done and been many things, but if they work for the military it is difficult to use the labels 'umpire' or 'referee'. There is no such thing as neutral mapping, whether it is being done on paper or from screen. From Cambodia to Iraq, whether one family died or another lived may have depended on how a modern-day American trained in cartography chose to interpret a few shades of light on dark, all in the space of the few seconds which the technology gave them to decide.

measuring the length of time between the first signal and the strongest signal in a sequence, navigators of German aeroplanes and airships could get an idea of where they were located in relation to a fixed point on the continent and hence could more accurately bomb London during the First World War. However, the signal was most useful for returning to Germany, as its accuracy decreased with distance (ironically it could be used as a direct route into the beacon by enemy aircraft): wartime navigation has never been simple. Chapter 3 covers cartography for navigation in more detail; here we are interested in how changes in navigation technology have altered cartography.

There are, at least, fewer things to worry about when navigating aircraft, particularly commercial airliners, in peacetime. In fact aeroplanes have been landed automatically for over three decades, using radio signals. At the time of writing, the systems are moving to using microwaves, but the basic idea, that it is usually safer to land a plane without the help of a pilot, has become widely accepted. What relevance does this have for mapping? Well, without the need for pilots, except to be there in case of emergencies, there is very little need for conventional maps, either on paper or screen. Apocryphal tales of pilots searching out school atlases to be able to locate their position abound. Five hundred years after the most well-documented crossing of the Atlantic by sail, today's navigators often have as little practical knowledge of how to steer a passage to the West Indies as Columbus had. The difference is that they now have computers to do it for them. Computers have not only helped us draw more elaborate pictures of the world; they may progressively make us more lazy about learning where we are, as we can rely increasingly on these machines to 'know'.

Most international passengers will now have a better idea of the geography of their national airports than of the countries which they fly over. They may well have spent more time waiting in airports and travelling around them, than in the air. Once they are in the air they, and the pilots, can see little but cloud, sea and sky for much of the flight (and most passengers do not sit by a window). Some airlines provide crude television maps on which a symbol of the plane is shown to be slowly moving. These often use Mercator's projection, even for near-polar flights, and they invariably show you little more than approximately what ocean or continent you are over. How has this disinterest in where we are arisen? One possible answer has to do with giving people a sense of security. For many there is no need to know where they are: the computers know and so we should not worry. Some air travellers certainly do not want reminding that they are in the air. As travel, through improved navigation, becomes easier, the world as we represent it in our minds changes shape. Two cities are linked by their airports and by an inconvenient but at least relatively short spell spent cramped in a metal tube (the aeroplane) travelling between them. As people have become more successful in mapping the world technically, their success has reduced the perceived need to map it physically. This process has now reached a point where, by the end of the twentieth century, it can be claimed that the most basic of ancient map users, the foot soldier, no longer needs a map:

> in 1991 single-seater pest-spraying aircraft in the Sahara desert were routinely fixing position by GPS with an accuracy of 30 m. In this they were ahead of the airlines, which were equipped at the time with VOR and DME and, where appropriate, Loran C and Omega [land-based radio beacon systems], which provided area coverage of a lower accuracy. Spraying aircraft, during the locust plagues in the same area in the late 1980s, had only their compasses and their maps to guide them over terrain which often lacked clear features. It is in such low-cost operations that the impact of GPS has been most immediate and most dramatic. The military share these serendipitous benefits. For the first time the infantryman has a universal fixing system which does not rely on a map, an application which can scarcely have been envisaged when satnav [satellite navigation] systems were first perceived as a possible solution to the nuclear submarine's navigational problem. (Williams, 1994, pp. 293–294)

The quotation above is a useful reminder of the diversity of uses to which new technology is being put. From tracking the age-old plague of locusts to guiding the world's travelling nuclear arsenals, the new beacons in the sky are ever present. In terms of visualization they negate the need for traditional maps showing features that are likely to be prominent on the skyline. If your GPS receiver tells you where you are and where you have come from then you are likely to have a very good idea of where you are going. Maps of the future may well serve less of a function of helping us to find our way round places; it is more likely that they will show us something about those places. Maps of the world no longer need to use a Mercator projection because the reason for that use, i.e. dependency on the compass, has gone (see Box 6.2).

Old and new views

Not only can new technology offer benefits to cartography in terms of new data and tools for representing the world, but past cartographic practice provides many lessons which have still to be learnt by those who are designing many of the 'new' techniques of scientific visualization. This is perhaps most true in the field of terrain mapping where cartographers have developed some of their most effective illusions to put their interpretations of 'reality' onto paper – turning a few brush- or pen-strokes (or spray-gun bursts) into a mountain range which both looks realistic (despite their being no true realistic image of the whole range to view) and usefully informs us about the structure of those mountains.

The drawing of terrain maps, which show the landscape from 'the side', as opposed to from 'above', which is the conventional mapping practice, mixes

Box 6.2   Contemporary mapping box – Mapping time travel

'The Path of Minimum Time' is the title of a chapter in Williams (1994). In it, Williams describes how the pursuit of minimizing travel time rather than distance only became a priority of sea travellers in the nineteenth century. Before then so little was known of currents and so much reliance was placed on the wind, that it made little sense to worry unduly that the shortest path might not be the quickest. It was with the development of air flight that the path of minimum time became of primary importance. In 1929, the (British) Meteorological Office and the Royal Air Force showed that the minimum time path from Bedfordshire in England to New Jersey in America was 40% shorter than the great circle route because of spatial variations in prevailing wind speed and direction. Although the Hindenburg disaster terminated the use of airships which may have otherwise flourished, the growth of North Atlantic air traffic after the Second World War led to a renewed interest in minimum time paths. By 1949, aircraft were saving a couple of hours in flight time across the Atlantic by following the minimum time path route rather than the shortest distance route.

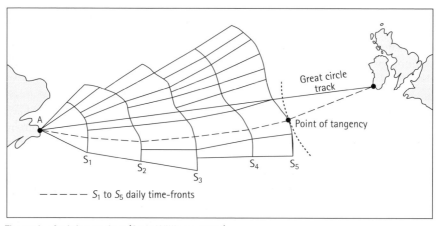

The path of minimum time (from Williams, 1994).

The importance of developments in air navigation for cartography was that minimum time paths vary and are calculated using data from ocean weather ships and from on-board instruments, whereas the great circle routes can be drawn on maps and are fixed. What is more, minimum time paths through the air are three-dimensional, so the idea of using a map to find a route by air appeared very outdated even by the middle of the twentieth century. However, only ten years after the savings from varying aeroplane routes had been calculated, the routes were switched back to the fixed tracks drawn on a map. This was because the congestion in the air had reached a point where it was no longer safe to let each aircraft navigator attempt to find the route of minimum time

(continued)

(continued)

travel. If they were successful they could end up colliding. A further decade on and the routes were again allowed to vary, now in response to day-to-day atmospheric forecasts and the volume of traffic anticipated.

The figure shows how a minimum time path can be constructed for a ship voyage from Ireland to the United States of America. This example is illustrative of maps that were drawn by the US Hydrographic Office in the late 1950s based on wave and wind forecasts. The more southern route can be seen to be the shorter because a straight line from Ireland reaches the fifth day line most quickly along it. There were only five lines on the map because forecasts for future wind conditions could not be reliably made for periods longer than five days.

skills from both science and the arts. Terrain maps are not sketches of the landscape and have always been accepted as maps in cartography despite their unorthodox perspective. Terrain maps present an illusion of three dimensions upon a two-dimensional medium. A number of techniques are used to achieve this. Vertical exaggeration is the most important technique. Without this even the most rugged mountain range appears to be almost flat. One reason vertical exaggeration is desirable is that people view vertical distance as far more important than distance across the ground: in our daily lives heights tend to be difficult to traverse, slopes are more difficult to build on, cliffs present dramatic landscapes – particularly from our usual viewpoint, on the ground. Automated three-dimensional mapping cannot simply construct perspective images of landscapes on the computer screen. To create successful images, procedures for exaggeration need to be considered and supplemented by the use of other techniques such as depth cueing and shading and the inclusion of artefacts such as shadows from an imaginary light source. What appears to be the most informative image may not be that produced by trying to reproduce the precise pattern of sunlight that ray-tracing computer software would reveal (ray-tracing is a computer graphics technique whereby, for every pixel on the computer screen, the rays of light which would theoretically emanate from it to the viewer's eye are traced backwards to their surfaces of reflection, refraction and eventually to their point of origin to determine what colour that pixel should be). Further representational methods such as rendering imaginary vertical poles, seen rising through a translucent landscape, can be employed (as one form of depth cue). However, since the products of many of these apparently sophisticated systems are viewed it becomes clear that for those currently designing systems to render surfaces through computer graphics, there is a great deal to be learnt from traditional terrain-mapping techniques (see Further reading and Box 6.3).

Past cartographic practices have tended to represent the complexities of real-world geometry in the limit, by surfaces, which can be thought of as two-and-a-half-dimensional objects. They exist in three dimensions, but unless they contain features such as cliff overhangs, tunnels or caves, then at every point on the plane there is only one height value. Such objects are relatively simple to use

Box 6.3   Personality box – Eduard Imhof and terrain mapping

Eduard Imhof was born in the Canton of Graubunden/Grisons in Switzerland in 1895. It was here that he gained his passion for mountains, becoming a leading member of the Swiss Alpine Club in his twenties. However, before then his family had moved to Zurich where he trained as a survey engineer before working as an assistant in the Geodetic Institute. In 1925, he was elected to the first Swiss Chair in Cartography and founded the Cartographic Institute in Zurich. He was also well known as the first president of the International Cartographic Association which he was instrumental in founding in Berne, Switzerland, in 1959.

Terrain mapping (from Imhof, 1982).

He developed his own style of mountain relief cartography in the 1920s which, 60 years later, formed the basis for publication of the English translation of his book, *Cartographic Relief Presentation* (published in German in 1965). This was just one of many books and articles he published over a very long career. In his eighties he was still active and the Swiss Federal Office of Topography published his new water-colour relief map painting of all Switzerland in 1982 (Imhof was an established landscape painter as well as

(continued)

(continued)

being 'the doyen of European cartographers'). In 1980, he was uniquely awarded the Carl Mannerfelt Medal of the ICA. Eduard Imhof died in Eilenbach in 1986, aged 92.

Imhof worked on perfecting techniques for terrain mapping in an era when the use of computers in cartography was becoming more and more commonplace. He expressed a certain scepticism over the utility of computer-aided mapping, reminding cartographers that the computer could not determine what maps should contain, nor determine the form of their contents: an assertion which, today, is even more important for those working in human cartography (an embryonic subfield of the discipline in Imhof's day) to consider.

The figure provides an example of the some of the types of mapping about which Imhof taught and which he created. The figure contains the Klausen and Pargel Passes in the West Glarner Alps and employs oblique hill shading graded by the application of aerial perspective (the scale of the map is 1:200 000). In the map the extent of the larger land forms has been emphasized to provide a structure to the mountains that the map reader can use to appreciate their form. The contrast between light and dark shades on the map is sharpened towards the summits of mountains and softened in the lowlands, more than it would be if this were a photograph of a model of the mountain range. This helps to stress the positions of mountain summits. More subtly, on the lighter side of major watersheds all local shadows are further lightened, and on the shaded side they are further shaded, to help emphasize the more extensive forms in the landscape – and thus to bring order into the picture of these mountains.

and comprehend. However, scientific visualization is often concerned with objects that are arranged in a truly three-dimensional structure such as the structure of a geological cross-section or a complex molecule. As cartography becomes more varied, representations of the physical and human geographies of the world are also more likely to include three-dimensional geometries. Reconstructions of vegetation using pollen analysis of borehole samples from a number of sites can result in a three-dimensional model where the third dimension is time and within which particular volumes of space can be coloured to represent the space–time extent of a particular species of plant. For example, the three-dimensional volume depicting the spatial diffusion of elm trees in Britain might exhibit a cone-like base, representing the geographical spread of this tree from a southern core following the last ice age. The cone would become broken up in its upper sections as the spread of clearing for farming destroyed much of the species. In the plane representing the most recent years there would be a plateau cutting off most of the volume following the species' decimation by Dutch elm disease. To view such a complex model of the changing vegetation surface of the earth, even for one species, would require the use of translucent shading (so that volumes could be viewed inside and behind other volumes); rotation of the three-dimensional space on computer screen (so that a visual illusion could be achieved of it as a real object); and finally the visualization software should allow particular two-dimensional slices of the volume to be viewed at will (such as the distribution of vegetation 2000 years ago or the plane which cuts through the maximal extent of a particular species' spread in both space and time). Similarly complex examples from human geography are not difficult to envisage, e.g. the spatial diffusion and collation of ethnic groups in Europe over the last millennium.

Another level of geometrical complexity is reached when considering four-dimensional mapping. The most common occurrence of this is when a three-dimensional object which changes over time. A simple example would be the position of the author's fingers while typing this sentence: this is the geometry of the world we live in. Real-world geometry can be considered three-and-a-half dimensional as in the real world we are unable to travel in time and so position in the fourth dimension is not a free variable. However, in representing and visualizing the real world there is nothing to stop us from moving through time freely. Take, as an example, mapping the movement of aeroplanes as they land and take off from a major international airport during one day. The representation is a four-dimensional object which might well contain many hundreds of individual elements representing the flight paths of individual planes. An animation of the day's flights, say from a viewpoint looking down on the airport is an obvious starting point to map its operation, but there are many other possibilities. It may well be that, viewed from a particular angle (e.g. from the south), the operation of the airport presents a very simple pattern (depending on the prevailing flight-paths). More complex viewpoints are equally valid, as are those which change in time and space (an example is given in Box 6.4).

Fractals: scale–free mapping

While dealing with objects existing in space which are more than two-dimensional, the search for simple patterns becomes more important because the human mind finds higher geometries, represented by higher dimensions, difficult to imagine. Fractals, by one definition, is the name given to objects whose dimensionality is not an integer number. Here, however, the story of fractals is told from a simpler beginning by considering another effect of the development of computer graphics; in this case visualizing satellite remote sensing data and viewing the world at different scales.

It has been shown how different satellites have changed the way it is possible to map from space and how navigation on land, sea or in the air can be undertaken. But how have they altered the way in which the world is seen? How they have altered people's views of the physical world is most easily appreciated, and much of this has been discussed above and in Chapter 3. How they alter views of the social world of towns and cities (which most of us inhabit) is less obvious. Satellite data are now being used to analyse patterns of land use and the design of cities. The effect of this is to begin to change the way people think of cities, most unusually perhaps, in terms of the 'structured randomness' of their design which is revealed through remotely sensed images. This is because satellite data have the advantage over conventional maps of not imposing the cartographer's

Box 6.4 Contemporary mapping box – The three- and four-dimensional mapping of disease

How might we search for patterns of disease within a population? There are numerous statistical methods developed to address this problem but almost all of them require an *a priori* knowledge of what patterns one might be searching for, before they can be found (clustering around points, for instance). To look for patterns without having a prior theory as to what we might find, we need to examine the graphical marks representing the patterns rather than analyse them. However, if we draw a map of a country and then plot the locations of people who have a disease upon that map we will tend to see a distribution that is remarkably similar to the original population distribution of the country. We could, of course, calculate rates of disease and then shade a map in to show the propensity of people living in different areas to suffer from a particular malady, but that requires the choice of areas to calculate rates for, which itself can distort the results. Both of these methods also ignore all the information we might have on the time when people were diagnosed to have these diseases.

A solution to disease mapping, made possible largely through the advent of scientific visualization, is the drawing of a dot map of the incidence of disease, but to do this upon an equal-population cartogram (see Chapter 8). If the disease is randomly spread across the population then the dots should form a random pattern because areas on the map represent true population on such a cartogram. More importantly, because we are visualizing these dots using a computer they do not need to be drawn in two dimensions, but can instead be placed with a volume, the third dimension representing time. Because we wish the dots (spheres in terms of their rendering) to be evenly spread in space if there is no spatial structure to the disease, the time-scale should be defined so that an equal number of lives are contained within any volume. The exact relationship between time and distance can be determined by looking at how far, say, the average person migrates in their lifetime and making this distance equal to the length that is used to represent the average lifespan in this space–time cube of human geography. Having constructed a volume of even space–time population density, each diagnosis of a disease can be placed within the volume and the volume viewed from any direction.

The most obvious direction in which to study a three-dimensional map of disease such as this is from the future looking 'down' on the map so that the most recent incidences of the disease are nearest and the oldest are pin pricks of light furthest away. From this angle north, south, east and west appear in their usual orientation. The virtual camera which is used to construct this image can then be swung down, say to the east at a time point mid-way in the time series of data. Then from this angle a vertically rising series of cases can be seen with the top of the picture being most recent and the bottom the most distant in time. North would be to the right and south to the left. The value of such an approach is that a pattern of cases which appears to form a cluster from one

(continued)

(continued)

angle (say in space) can be seen not to form a cluster from another angle (e.g. they are too widely separated in time). A disease that is transmitted from one patient to another by near contact will tend to form a trail of points within the space–time cube, while unconnected cases should not have a pattern to their collective geometry.

sense of order on the world. They also have the advantage over aerial photographs of, first, covering a larger area, secondly sensing a wider range of the spectrum (for instance by detecting heat) and, thirdly, automatically providing information in a form that is ideal for analysis with a computer.

That the information gathered by satellites is automatically produced to be analysed by computer does not imply that the analysis is simple to perform, nor that this is necessarily an appropriate way to study human societies. For those interested in the complexities of ascribing particular pixels to particular types of urban land use, Mesev *et al.* (1995), from which the example here is taken, should be consulted. In essence, it is possible to distinguish between different types of housing from the densities of building material used in their construction – the amount of land they cover in concrete, brick and slate. Population census data can be used to 'train' the procedure being used to classify satellite data so that they produce pictures which conform to part of our quantitative knowledge of the real world, but provide more detail than the census, whose data are aggregated to preserve confidentiality. These detailed pictures can then be used to study aspects of the pattern of cities which can only be analysed with high-resolution data (such as their fractal dimension – see below) and to study the change in cities at a much finer temporal resolution than can be done with decennially collected census data. A more mundane use of this information is, of course, as a basic data set for drawing more detailed maps without the full costs of surveying.

What do we mean when we say that the structure of a city might be fractal in nature? 'Fractal' was the name given to objects which could be described by mathematical functions but which are not, to put it simply, 'smooth' (Mandelbrot, 1983). However, this does not mean that their shape is a random mess. There has to be a pattern in its irregularities for an object to be considered to be fractal. What is more, that pattern has to repeat itself at a variety of scales. Why are such objects interesting? Apart from the fact that they are often quite beautiful, it is because they can be described by what initially appear to be very simple mathematical equations. This is of interest to scientists because much of science is about the search for simple laws to explain complex outcomes. In social science, from where the examples used here are taken, the complex outcome might be the shape of a city – one part of its urban form. It has recently been suggested that a new way of trying to understand the evolution of cities is to see them as fractal objects (Batty and Longley, 1994). This is of interest to cartography because so much of cartography involves the mapping and hence the generalization and simplification of cities. But if we are trying to simplify an object whose inherent nature is complex, are cartographers not fighting a losing battle?

The most simple (and for some purposes the most important) object shown on many maps is the boundary between sea and land. From world projections to the largest-scale mapping, this distinction is seen as essential. Yet if we consider this boundary carefully it cannot be drawn correctly. Unlike a national boundary drawn as a straight line on a map (which is how many were created), or the edge of a road or building which is relatively straight, a coastline can always be drawn more 'correctly' if more detail is added. This is, of course, to ignore the problem of tides and the fact that the coastline moves over longer time periods. What is interesting about the cartography of coastlines is that if someone was shown a map of part of the west coast of Scotland, or one edge of a lake in Vietnam, or a section of the bank of a river running slowly through some hills in Chile, they would have a great deal of difficulty telling which each was unless they were already familiar with these shapes. Coastlines at a wide variety of scales tend to look the same and so cartographers have been able to develop a variety of techniques for dealing with these objects which are largely scale-invariant, and which have been applied in objective methods of line generalization (see p. 40). It may well be that the fractal aspects of many natural objects (i.e. the repetition that exists in the irregularities) helps make their simplification on maps and subsequent recognition by map readers easier. The more that is understood about these processes, the more imaginative our mapping may be.

As Batty and Longley (1994) have demonstrated with cities, it is not only the natural objects which are mapped that exhibit fractal properties (see Further reading). Take, for example, the shape of Greater London and the smaller urban areas of its hinterland (Figure 6.2). The smaller urban centres can be seen as 'tiny Londons' in their own right, while the irregular boundary of London itself produces enclaves and exclaves which appear to be repeated many times around its perimeter. What is remarkable about this ordered randomness, is that London is supposed to have been one of the most planned and controlled cities in the world during the period of its latest expansion and yet what has been created appears, at first sight, to be a mess. With more thought it can be argued that the same sorts of processes that constrained and encouraged the growth on its western side operated on its east, and so it is not surprising that there is regularity in the irregularities. Enthusiastic developers might have managed to break out into a notional green belt at one point, with the results that planning regulations were subsequently tightened up around that area, resulting in the spur and enclave pattern that is so familiar to the boundaries of cities (with the spurs usually following major transport routes out of town). However, one result for cartography is that a simple stylized blob to represent London may well be an unnecessarily simple distortion of its true nature.

Notwithstanding their common spatial nature, to suggest that human and physical phenomena can be mapped in the same way is currently a contentious viewpoint to take within the study of geography. Much that divides the discipline at present revolves around debates over the use of different techniques to study different subjects. In general, there are valid reasons for these tensions: the

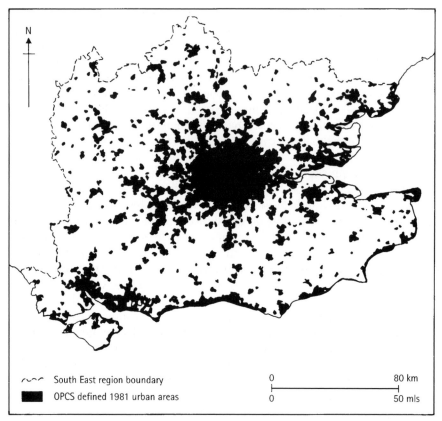

Figure 6.2 The fractal boundary of the urban area of London (from Batty and Longley, 1994).

study of forest soils and the study of socialist societies are a world apart, and yet both are part of geographic enquiry today. As we argued in the Introduction to this book, maps and mapping form a bridge between different fields of geographical study. This is not to suggest that a knowledge of, say, soils is useful in the study of political geography (although, ironically, there are close connections between these aspects of geography because industries associated with mass labour were originally located near particular mineral resources, which have influenced soil type). Rather, we argue, the cartographic skills required to design a map of 30 soil types and to decide whether or not to show mixtures of soils in explicit transition zones, are very similar skills to those needed in the mapping of spatially complex political allegiancies. The products of this cartography can also show similarities in patterns which occur despite the subject chosen. For instance, within an area of soil type A in a country there will be patches of soil type B which themselves contain areas of type A, and so on. The same can be shown to be true of the geography of voting in a country such as Britain. Maps generalize, but they also deter the map reader from making even more sweeping generalizations. The recurrent patterns of one group lying within the confines of another are, of course, fractal.

The recognition of fractals generates solutions as well as problems for cartographers. Simple algorithms can be written to literally 'wiggle' rivers, so making the map appear more 'realistic'. An old trick taught to American schoolchildren who had to draw a map of Britain quickly was to draw a thin triangle with its point at the top, but to wiggle the pen as they did this. All that is now being done with computers is a more sophisticated version of this trick, with a few more rules. For instance, a computer-generated map of Britain could represent London as a circle, but perturb its boundary using a simple equation to make it appear more random and 'true to life'. The advantage of using the equation is that it will continue to work at a variety of scales. Thus, as the viewer zooms in on London in this map, parks would appear within it and the urban boundary would appear to be even more detailed, as it is in reality. The only problem, of course, is that the parks would not be in the right place! What then is the use of such technology? Well suppose now that a map of an American city was being viewed and zoomed into. The map could appear to be more realistic by drawing, within each land parcel, a house, each of which was a little different (unlike the uniformity of British homes) from every other house. More obvious perhaps, a map of a forest could include trees that were not all shown by the same symbol. Maps of usually dry stream channels on a mountainside could be shown more realistically by using a dendritic pattern than by using an overall symbol for 'there may be streams here'. Is this really cartography? Perhaps not, and it can be a little dangerous, but then cartographers have always employed artistic licence in their work (as the discussion on terrain mapping above exemplifies): fractal techniques can be seen as simply a more scientific method of doing that.

Fractal geometry can be used to summon up more than just a realistic image of a forest from digital map information which says that there are trees on a particular slope. It can be used to generate artificial slopes themselves and realistic pictures of the mountains and valleys which they form, the rivers which run down them and the city settlements which are spread along the banks of those rivers. Eventually an entire island can be grown *mathematically*, and this island can sit in a sea of islands around a fractal continent on a fractal world in a fractal galaxy, and so on. A few simple equations, whose use was only first recognized with the development of computer graphics, can today be used to build planets which appear to be almost as realistic as our own and which have their own cartography in the form of the pictures which are drawn of them. There is no longer one world (or even a few worlds) to map.

Further reading

S. Hall's *Mapping the Next Millennium* (Random House, New York, 1992) is a key introduction to how cartography is currently evolving from the viewpoint of science and computing. J. E. D. Williams' *From Sails to Satellites* (Oxford University Press, Oxford, 1994) fulfils a similar role, but from the standpoint of the evolution of navigation. The first edition of J. Gleick's *Chaos: Making a New Science* (Heinemann, London, 1988; 2nd edition, Abacus, London, 1993) was one of the first popular books on the new science of uncertainty, and is well worth returning to. However, to see the origins of much of the work on what is new in science the original report on *Visualisation in Scientific Computing* by B. H. McCormick, T. A. DeFanti and M. D. Brown (eds) (Special Issue of *Computer Graphics,* **21** (6), ACM SIGGRAPH, New York, 1987) should be read. There is a synopsis of this report in *IEEE Computer Graphics and Applications,* **7** (1987), 61–70. For an alternative view, see J. R. Beniger and D. L. Robyn, 'Quantitative Graphics in Statistics: A Brief History', *The American Statistician,* **32** (1978), 1–11.

T. Rabenhorst and P. McDermott's *Applied Cartography: Introduction to Remote Sensing* (Merrill, Columbus, Ohio, 1989) is well illustrated and takes the reader through the basics, step by step. An alternative image of some of the early uses of remote sensing is given in the introduction to William Shawcross's *Sideshow: Kissinger, Nixon and the Destruction of Cambodia* (Hogarth Press, London, 1986). *A Guide to Remote Sensing: Interpreting Images of the Earth,* by S. Drury (Oxford University Press, Oxford, 1990) gives a useful introduction to remote sensing applications.

On terrain mapping, see A. G. Hodgkiss, 'The Bildkarten of Herman Bollman' in *Canadian Cartographer,* **10** (2) (1973), 133–145, and M. Wood, 'The Panoramic Map of Central Scotland', *Bulletin of the Society of University Cartographers,* **17** (1) (1984), 1–7. E. Imhof's *Cartographic Relief Presentation* (Walter de Gruyter, Berlin, 1982) provides an excellent guide from a well-acknowledged master of this art (for a brief description of Imhof's achievements, see the address of F. J. Ormeling Snr in *The Cartographic Journal,* **24** (1) (1987), 83).

On fractal geometry and the interpretation of remotely sensed data, see T. V. Mesev, P. A. Longley, M. Batty and Y. Xie, 'Morphology from Imagery: Detecting and Measuring the Density of Urban Land Use', *Environment and Planning A,* **27** (1995), 759–780. On fractal images of nature in general and cities in particular, see B. Mandelbrot's *The Fractal Geometry of Nature* (Freeman, San Francisco, 1983) and M. Batty and P. Longley, *Fractal Cities* (Academic Press, London, 1994).

Chapter 7

Geographical information systems

Introduction

Although the world is awash with maps, fewer and fewer cartographers are drawing these pictures. The artistic mystique of the scribing pen and set-square have been replaced by the scientific black-box of the digitizer and the geographic information processing engine. Computer scientists now control the mechanics of map design as well as map production. Because of this, students today are far more likely to take a course in GIS (geographical information systems) than in cartography and are likely to know more about polygon topology than topographic hill shading.

GIS have their widest implications outside of the classroom. Inside the computer and in the minds of some analysts the world has become a 'coverage' or blank canvas upon which the information which it is expedient to collect can be draped. A generation of students is being trained to learn, using contemporary spatial data handling techniques, how to better manipulate the world: how to discover and develop markets; to identify and eliminate clusters of crime or disease; to assess the environmental impact of a waste dump. Through these systems a new representation of the world is created and a new way of engaging with the world is developing. In this chapter the history of GIS is summarized, their current state analysed and their future debated.

A short history of geographical information systems

Geographical information systems are primarily commercial products and so it is apt to define them using the words of the one person who may have invested in and profited most from this software: 'GIS is an organized collection of computer hardware, software, and geographic data designed to efficiently capture, store, update, manipulate, and display all forms of geographically referenced information' (Dangermond, 1992, pp. 11–12).

Various incarnations of GIS have been available for over three decades but they have only become commonplace in a variety of organizations in the last few years. For our purposes here, the histories of GIS used for commercial practices and for academic research are similar enough to treat them as following the same trajectory. For all systems over this period the software has become far more sophisticated and there is now a wide range of products available. However, the fundamental components of these systems have not altered over time. A brief

Box 7.1   Personality box – Jack Dangermond and the radical view of the world of GIS

Jack Dangermond is credited with creating the first successful GIS company, Environmental Systems Research Institute (ESRI), after he graduated from Harvard in 1968 (see also Box 7.2). Ironically, the company began as a not-for-profit organization and launched its main applications package, ARC/INFO® in 1982. This product is now the most successful GIS software package on the world-wide market and the company is certainly now profit-making (Coppock and Rhind, 1991, p. 32). Given the global spread of GIS, Jack Dangermond has developed a particular view of the capabilities of the software and the people who use it:

> GISs now perform the function of data integration machines for many kinds of data. . . . This is an extremely important function of GIS technology, perhaps the most important function it performs, since it provides one mechanism by which people in different organizations, different levels of government, different countries, different disciplines and even with differing political views and goals, can come together around a common resource to share the process of solving common problems'.
> (Dangermond, 1992, p. 16)

Jack Dangermond.

Although it is difficult to imagine rival warlords splitting their differences as they pore over the electronic map (unless they became so confused by the GIS that they forget what they were fighting about!) it is not difficult to find evidence of these systems fostering co-operation between different academic disciplines or different tiers of government. Their databases require and consume such large quantities of information that it is very difficult to make any progress without substantial co-operation. The extent to which this enhanced level of 'coming together' actually leads to the solving of common problems, and whose problems these are, is where the source of greatest debate lies (see p. 125).

Jack Dangermond has a vision that GIS can help in the development of an 'electronic democracy' by helping to establish links between the citizen and the

(continued)

(continued)

increasingly remote institutions of government. New technological developments will aid this project, brought about by fibre optic communications,
multi-media systems, cheap hardware and better database management
systems. The citizen should be able to easily access the information which
institutions hold and when it can be geographically disaggregated they will find
the information engaging, because it relates to where they live. Whatever the
merits of such arguments, it is interesting to note that these were made at the
International Cartographic Association conference in 1991, some time before
the most famous proponent of such ideas, Bill Gates (who started the Microsoft
Corporation), became famous for them.

description of a GIS would include a software system which incorporates programs to store and access spatial data, programs to manipulate those data and
programs to draw maps. Because of this structure, the origins of modern systems
can be found in a combination of data management programs, computer-aided
design (CAD) programs for automated mapping (which became know as
Automated Management/Facilities Management (AM/FM) systems), and early
geographical information processing systems. A geographical information processing system is a set of programs which input and output spatially referenced
data, transforming them in some way in the meantime. Many geographical
information systems are still amalgamations of data analysis and data management programs, the most famous example being ARC/INFO® which is a combined geographical information processing and database system (hence the
two-part name of this product).

   In terms of ways of representing the world, GIS can be viewed as a new technology with an impact akin to the introduction of the microscope or telescope.
The technology makes visible a previously unseen perspective, opening up new
worlds to our eyes. Some of these new worlds are macroscopic vistas of hugely
detailed maps; maps which could have never been drawn by human hands, such
as the 1 km^2 resolution satellite image reconstruction of the vegetation cover of
the United States of America. At the other extreme, GIS can create new microscopic images: for example, which is the street with the most wealthy people in
your town? This can now be determined from the intersection of land registry
and spatially referenced share ownership records with a plethora of other private
databases. The revolution is not in the availability of satellite images or individual level databanks: it has been in the development of software to analyse
these sources.

   An alternative view of these systems is that they are simply the digital development of cartography combined with a great deal of sales hype. Proponents of the
technology have responded to claims such as this by arguing that GIS will super-
sede cartography in providing the dominant world-view. They argue that the
paper map is obsolete and that a new digital, constantly updated, effortlessly
browseable medium is being created. They add that there is nothing particularly
invidious in these developments which, in one sense, simply extend what human

Box 7.2   Contemporary mapping box – Earliest geographical information systems

The earliest major implementation of a geographical information system was by the Canadian government and was designed for managing forestry and other types of land use. The Canadian Geographical Information System (CGIS) was designed in 1966 by a team led by Roger Tomlinson, a geography PhD graduate.

In different parts of the world different professions were associated with the development of these systems and this has contributed to a wide variety of flavours to GIS: for instance, surveyors have dominated development in Australia, local government is important in Britain and the utilities (including natural resource areas such as forestry and mining) matter much more in America. What was common in the diffusion of these systems was a bias towards affluent English-speaking countries, such as those just mentioned. This has resulted in what could be viewed as American 'computational imperialism' in terms of the diffusion of software (see also Box 7.3).

There are many other contenders for the title of earliest GIS, depending on how development and implementation are defined. As with the Canadian example, key personalities have been allowed to emerge as the history of each research group has been written. The Laboratory for Computer Graphics, Harvard University, was established by Howard Fisher in 1965 and it created the SYMAP package, initially for producing maps on computer line-printers. Although SYMAP was one of the earliest packages to be produced, it was not successful for long as innovation stepped up a pace, and the Harvard laboratory went on to spawn some of today's most successful products. Both the Intergraph and the Environmental Systems Research Institute (ESRI) companies were created by former graduates (see Box 7.1).

A tale of pioneering figures, similar to those guiding American developments, can also been told for Britain. Just as Jack Dangermond was leaving Harvard, David Bickmore was establishing the Experimental Cartography Unit at the Royal College of Art in London. The private market for digital information in Britain is severely restricted by issues of copyright and so it was the government-funded Ordnance Survey which benefited most directly from collaboration with this research centre (Rhind, 1988).

What is interesting about these examples is not necessarily the list of individuals involved, but the coincidences in their progress. Both academic research centres flourished in what might appear to be inhospitable environments: a department of landscape architecture (in a university without a geography department) and a college of art. It may be that innovation is more likely where research conditions are uncomfortable! The timing on both sides of the Atlantic, and with the CGIS across the Canadian border, was almost identical. All these developments were technology-driven as the appropriate computers arrived in each research institute at roughly the same time.

beings have always done with maps. What may be more invidious is the message, inherent in much writing on GIS, that if we only had more data we could draw a truer picture and that, eventually, with enough data, all will be revealed. This way of thinking last held sway at the turn of the century when it was thought that, with complete information, key moral and political problems could be solved through the search for an optimum solution.

A further set of retorts to the claims that GIS are particularly revolutionary can be found in both technological and theoretical arguments. Technically the systems can be seen as particularly clumsy examples of a style of programming popular in the 1960s and 1970s. The size of the code which runs many of these packages is disproportionately large and many of their algorithms are extremely inefficient. Theoretically the 'findings' from much GIS research appear particularly naive to many other researchers. They are likely to say 'we always knew the vegetation cover of the United States of America was so diverse', or 'of course that street contains the most wealthy people in your city, ask anyone!' Attacks closer to the heart of GIS often also involve the reliability of these systems: 'how can you be sure that that pixel is conifer forest?' or 'but the richest people are likely to hold their wealth in trusts, so how can you locate them?' Nevertheless the systems are now firmly embedded in the teaching and practice of representing the world and so the claims of their makers need to be considered in detail.

The democratization of map-making: removing the mystique _____

One of the most positive ways in which GIS can be viewed is as a process of democratizing a medium. Just as the introduction of the printing press allowed an explosion in the quantity of books in the world, the geographical information system has easily resulted in more maps being 'drawn' in the last decade than were created in all previous human history. Most importantly, these new maps are being created by a far wider range of people. The mystique of map-making of previous years (see Chapter 1) has been partly eliminated. Cartographers, who used to have the status of scribes, often now feel reduced to tidying up those maps which other people cannot find the time to work on. Meanwhile students today can, in a week, fill their dissertations with digitally produced illustrations that would have taken their teachers years to produce when they were students. Increasingly, as systems become more diverse and cheaper, map-making packages (which contain many elements of a GIS) are finding their way onto millions of home computers. How will the world be represented by people who have never been trained to draw it in any particular way? In one sense we may be returning to a world before formalized cartography, where many different people around the world developed their own, slightly different forms of mapping. The formalization of cartography created great conformity (see Chapter 3) – a conformity which is now only maintained by the machine. In other words, the contemporary restrictions of different mapping packages (they do have some representational and data-handling limitations) result in common standards today; in the past common standards were maintained by teaching the 'right way' to map. As the computer packages begin to allow more varied forms of map to be drawn, conformity is likely to decrease.

Exploring spatial data with a GIS

In order to explore this Utopian vision of the democratic map we have to dissect GIS into its basic components to understand what kinds of maps can be created by these systems. The first thing to realize is that the core of GIS is a piece of computer software, a complicated set of instructions. This software controls various pieces of hardware ranging from storage devices (disk drives and virtual memory) to central processing units (CPU, consisting of the silicon chips which process the data) to display devices (screens and printers). The software tends to be large and expensive to purchase. More importantly, it often requires the most advanced and expensive hardware to operate effectively. This is because spatial data holdings need to be extensive to characterize adequately the view of reality they embody. Large disk drives are needed and a great deal of memory is required for rapid-access storage; while fast processors and high-resolution output devices for display are almost always essential. All of these requirements greatly add to the cost of most systems and complicate their operation, as somebody has to know how all these parts operate to keep the system running. These factors have been largely responsible for curtailing the most widespread use of GIS. There is, for instance, no 'free' system bundled with an off-the-shelf computer as is often now the case with word processors and spreadsheets. Proponents of these systems might claim that this is only a transitory phase and reflects how powerful future GIS will be, but the reality – that many schools, for instance, currently cannot afford these systems – suggests that democratization of mapping still has some way to go.

It is also vital to realize that even the software and hardware together do not constitute a GIS. The most important component of these systems is the data they operate on (see Box 7.3). This is also often by far the most expensive part of the system, costing many times more than the software and hardware combined to purchase, collect, correct, manipulate and maintain (i.e. revise). Unlike a telescope where, after the initial purchase, all that is needed is a cloudless night, a GIS requires its data to be handed to it on a plate, disk or tape.

In most countries outside the USA the basic topographical and census data are copyright protected; the copyright is owned by the data collector (usually the state) and has to be purchased. Even where data are supposed to be free, the mechanics of getting the data into the right format have created a profitable private industry for data bureaux (see Box 7.4). However, most frequently, users find that the data they need cannot be downloaded or purchased. They have to collect the data themselves and this can involve labour-intensive work ranging from field surveys to the drudgery of digitizing (converting an existing graphical map into digital form suitable for handling in the CPU). Even when the information is free, such as with some (usually outdated) satellite data, the work required to correct and classify it is expensive to fund. Finally, there are the costs of updating and maintaining the data. In very little time these costs (the alert

Box 7.3   Contemporary mapping box – Ways of owning the world

There is no such thing as a free data set. Even public domain data have to be located and shipped to the user. The nearest to free data which exists is information put on the World Wide Web, but to access that you have to be able to be on the Internet (still not easy in most countries of the world: see Chapter 4).

There is, however, a large and growing amount of very cheap data. The largest source in the world is the American government which has decided that almost all the information it collects should be put into the public domain and supplied as cheaply as possible. Thus, a British school student can download the American census, some boundary files and a digital terrain model (put on the Internet by another government agency) and, say, draw some maps of what types of work people living in different valleys in the Appalachians do. That student could not do the same for the population of the Pennine Hills in England.

In much of the world outside the United States of America most government data are copyright and an even larger amount are secret (often termed confidential). In Britain, the Ordnance Survey claim a copyright interest in almost all the maps of that country and will lease the digital information they hold to users for a high charge, often only on an annual basis. The census information is also confidential and members of the public have to buy it if they want access (local government officers, health administrators and university researchers have it bought for them *en masse*). To return to our fictional school student with an interest in the north of England, a detailed digital terrain model (DTM) of the Pennines is held at extremely high resolution by the military, but it has been considered to be against the national interest to release this (it is, however, now possible to buy a 50-m resolution DTM of the whole of the UK). This kind of restriction is widespread in data-rich countries, but is not universal. In Sweden, for instance, the government allows anyone to access certain financial information about anyone else, information which would be considered highly confidential elsewhere (in Britain there is not even a question on income in the census). What data are available to use in a GIS is very much dependent on who you are and where you are.

The clearest example of the importance of data in GIS is the almost total dearth of studies which use GIS in the 'Third World'. In many of these countries the data which would be needed for analysis by GIS have not been collected: there are often more important things to be doing. Even if the data were collected, there is usually not enough money to buy, maintain and staff the machines to analyse them. Where this lack of information becomes most apparent is in world mapping, where huge holes in the maps are commonplace. Even a simple map of world population drawn in 1995 had to exclude countries in the centre of many continents (e.g. Bosnia and Rwanda) and the countries that are excluded are often those in which there is most interest. Geographical information systems can only reflect a little of the information they are given; they will not perform miracles.

Box 7.4   Contemporary mapping box – GIS and jobs

The earliest GIS (see Box 7.2) was implemented due to a calculation made in 1965 that it could save Canadian $8 000 000 by relieving 556 technicians of 3 years work (Coppock and Rhind, 1991). Almost 30 years later these systems are still advocated by official bodies because 'now GISs are making it possible to replace time-consuming and labour-intensive activity' (Her Majesty's Treasury, 1994). It is doubtful that the introduction of these systems does actually result in fewer people being employed in many firms and offices. Geographical information systems are quite labour-intensive operations, particularly in their requirement for routine digitizing and editing work. What is interesting here is that it has often been a desire to shed labour which has resulted in this job creation. It is not only routine employment which has been created by the demands of these systems. Across the world thousands of students have graduated from MSc and PhD programmes in GIS and many of these are going on to be paid to teach others about these systems. Unlike many other forms of automation (such as of bank tellers by automatic teller machines), GIS tries to automate something so complex that it often results in creating an even more complex solution. Simultaneously, the insatiable appetite which these systems have for data is leading to the creation of numerous data bureaux which collect and sell this information, again creating more jobs (but see Box 7.3). Jobs have been lost as a result of the introduction of GIS, most often in traditional cartographic operations, but it would be difficult to claim that fewer new jobs have been created.

reader should see a pattern emerging here) can be greater again in both time and money than those required to set the system up in the first place. Updating data costs much more than acquiring them in the first place, which often costs more than the hardware needed to run a GIS, which is quite likely to cost much more than the actual software itself. These systems are not cheap to operate, and this is before the costs of training and staffing are considered. A rough set of current ratios for total GIS costs are: hardware 1 unit; software (including training and updates) 10 units; data 100 units. Over time the cost of hardware is still falling while the cost of data continues to rise.

Modelling the world using GIS

Geographical information systems turn out to be an expensive form of democracy, but this may be only a temporary set-back. It is what these systems can do with the data that is most important to the argument that they can demystify the making of maps. In essence, some GIS hold their data as a series of layers, and the creation of new information is achieved through the construction of new layers derived from the processing of one or more of the existing layers. For instance, a layer of roads (represented as lines) could be used as input into a process which output a layer of road corridors (represented as polygons 'buffering' those lines). More demanding, perhaps, a layer of heights (represented as a digital elevation model) could be used as input to a process which output a layer of streams (represented as lines). The cartographic possibilities of adding value to existing data through the use of GIS are not difficult to imagine.

Two, usually mutually exclusive, models of the world have been used in most GIS to date and these have important implications for the way in which the world is represented and depicted using such systems. One is the vector-based, geo-relational, model in which elements of the world are represented by a series of points that, when connected, form lines which, in turn, enclose polygons. Such a representation is close to how cartographic information has been traditionally recorded, due to the methods of having to accurately reproduce, re-scale and re-project maps in the past using manual methods of drafting. There are a number of advantages to this model of the world, not least being the relatively simple and efficient (if somewhat tedious) process of converting paper maps into digital form using vector digitizing. The disadvantages have tended to arise from the fact that this is not the usual way in which information is stored in computers and so the manipulation of such information can be limited and slow. For instance, when reproducing maps by hand the generalization of features tends to be done simultaneously and subconsciously. Automating this process with a vector digital model is far from simple when using conventional serial-based computers.

The other model is a raster, layer-based structure in which a series of matrices or grids is created, each completely covering the area being mapped; the cells of the grid show the value of a feature at each point. Information such as height, or the presence or absence of a particular species can be efficiently stored by this method and subsequent manipulation by encoded algorithms is relatively simple. However, there are also disadvantages to this model: people tend to cluster in certain areas (cities, for instance) so that when mapping features of human society, the raster data model uses a great deal of computer storage space in recording 'empty' areas and very few of the data are being used to represent information about most of the people on the map.

In recent years more complex and combined data models have been developed to improve realism, storage and manipulation. One of these, developed from the raster representation, is the quadtree, in which space is recursively subdivided so that areas in which there is more detail have more storage space allocated to them. Another area of development, the antecedents of which can be found in vector models, is the object-orientated representation of real-world features (see Further reading for more details).

Figure 7.1 shows some of the basic data-structures and processing operations of these systems. Although this way of digitally representing the world might look simplistic, it allows an enormous range of products and conclusions to be developed. For instance, using GIS you can shade a map with the synthetic light of a setting sun to highlight the westward sides of hills; you can fill the roads with imaginary traffic based on the topology of the transport network and the distribution of population; or you can allow the viewer to zoom out from the image of a street to see its context with a nation, in seconds. However, as these

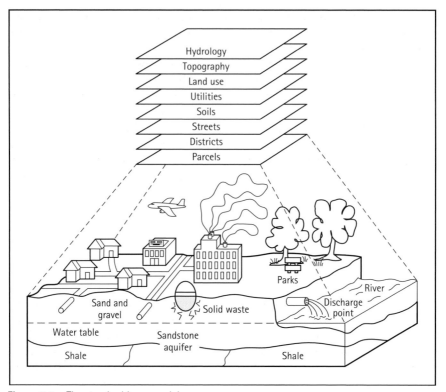

Figure 7.1 The standard layer model.

brief examples illustrate, there can be as much artistry and guesswork in digital mapping as ever existed in traditional cartography. In the real world, the streams will not be where the run-off model places them, the sun will not break unhindered through the clouds to light the westward slopes, and the transport system will certainly not work as smoothly as the theory suggests. These systems do not mirror the real world but, just like maps, they create simplified and idealized images of it. This is as true for the more complex data models as for the simple ones, although the more complex digital representations tend to produce pictures that *appear* more real.

Displaying 'world-views' using GIS

The final aspect of the technical operation of geographical information systems covered here is often that of greatest interest to map-makers: display. In relation to cartography, the greatest difference between GIS and traditional mapping has been the development of the 'virtual map': a map which is changing so quickly that any incarnation only exists for a few seconds or, in the case of animation, for a fraction of a second. The virtual map exists on the computer screen. By touching a dial the map-maker/map user can scan over the map in any direction, zoom in or out, alter the elevation and azimuth (if the image is of a surface) and, above all else, change the nature of the information contained in the map. They can choose whether to see the roads, how many roads to see, how the roads should be depicted (e.g. with width proportional to the extent of pollution from a model of their traffic). Then they can choose what else they wish to view alongside these roads: hospitals, rare plants, the index of local wealth created in our earlier example, or whatever. The extent to which this is possible depends on the amount and nature of the information the system holds, its technological sophistication and the imagination of the user, but in theory the manual world in which a few people made even fewer maps and from which the rest of us learnt our geography is being transformed into a world in which many of us will learn very different geographies – the ones we (or at least a large number of us) choose to see.

The applications of GIS and their effect on our image of the world

Geographical information systems are almost never explicitly used to alter people's views of the world. The most common use given and exemplified in the literature is to solve a set of apparently trivial problems: how best to route emergency vehicles to accidents, how to locate the most 'prominent summit' in a mountain range, or how to decide where children are unduly likely to be ill, for example. When these systems are then used to create images, the aim is often to create a map that looks as if it could have been drawn by hand, albeit over several years! In the world's best-selling system this desire to replicate traditional cartography means that maps are drawn on the screen as if a pen were being used (one line at a time, but in quick succession – very much a traditional vector system). Despite this general paradigm, even researchers within this field are often quite critical of the software they are using: 'Although the principles of designing displays of spatial data have been investigated for centuries, very little use has been made of such principles in GIS' (Buttenfield and Mackaness, 1991, p. 427).

The bulk of mapping done by GIS produces mundane choropleth or topographic maps as output, simply quicker than by hand. Choropleth maps are maps in which areas are shaded to depict the value of a certain variable (or variables) there, e.g. the political party for which most people voted for in each area. However, there are several clear areas where the introduction of these systems has resulted in the creation of new views of the world. In this section just two examples are given, one from physical geography and one from human geography.

In the geosciences in general, the three-dimensional display of subjects such as geological or hydrological data is of great interest. The introduction of GIS into these areas has developed from the automated creation of contour maps and simple isometric views of a surface to the interactive construction of 'fence diagrams' (Figure 7.2) and the use of translucent shading to allow hidden objects to be visible in the two-dimensional picture of a three-dimensional scene. These techniques have changed the way in which many researchers think about the

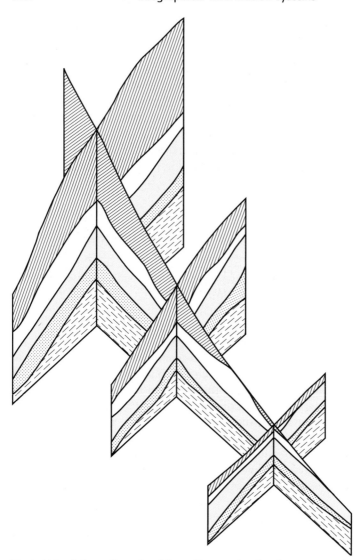

Figure 7.2　A three-dimensional fence diagram showing the structure of geology in a volume.

world, with further applications of such 'terrain modelling' beyond the geosciences, ranging from urban renewal to battlefield planning (McLaren and Kennie, 1989). Often the display devices for these systems can be stereoscopic (creating a slightly different picture for each eye) which, when viewed properly, help the viewer to imagine that they are seeing 'real' three-dimensional objects. Shading, shadows and speed of display are three other important ingredients necessary to sustain the illusion of reality that is required to allow the viewer's imagination to work on the problem at hand, rather than concentrate on the view.

In the social sciences the greatest impact of GIS in shaping people's perceptions of the world has come from geodemographic research, often initiated by

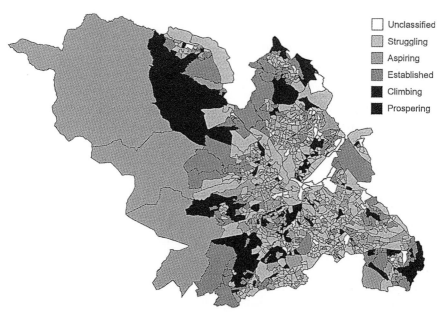

	Unclassified
	Struggling
	Aspiring
	Established
	Climbing
	Prospering

Figure 7.3 Geodemographic groups, by enumeration district, in Sheffield (note districts labelled as 'struggling', 'climbing' and 'prospering').

human geographers. Geodemographic research uses census and private data about the populations of people living in small areas to create profiles of those areas which purport to usefully describe the 'average' characteristics of the residents living there. The main product of this work, demographic profiles of particular locations and the people who are assumed to live there, has been for the business of market research and it is often partly blamed for the explosion of junk mail around the world. Cartographically, however, it has resulted in the production of some extremely detailed maps of the social structure of cities which, to an extent, will have changed the way that people, ranging from estate agents to students, view cities (Figure 7.3). Some of the most persuasive evidence that geodemographic mapping does affect perceptions is the condemnation of this work by other researchers.

Geographical information systems are making the world more spatially aware and making the map more commonplace. Most examples of their use involve the automation of what were hand-produced and paper-based products. Automated navigation systems (see Chapter 3), in which a console on the car dashboard shows a moving map of the road, are the natural extension of road atlases. However, over time the systems are beginning to evolve differently as researchers run out of manual products to automate. The automated navigation system appears to look less and less like a road atlas. The maps nowadays supplied on CD-ROM stop being scanned copies of paper versions and begin to take on a life of their own, by allowing for user interaction, by connecting with other means of presenting data (e.g. charts, tables, images, text and sound-bites) and by introducing novel methods of presentation beyond two dimensions, to include terrain views and dynamic sequences. The introduction of the computer into cartography has made a dramatic transformation possible in the way we represent the world. However, in general, people remain quite reluctant to look at the image very differently from the way they did before. It is still possible to walk into an international sales conference on GIS, look up at the banks of monitors stacked six high on the auditorium opposite and see hundreds of rainbow-coloured rectangular world map projections, the creation of which dates from the sixteenth century. The colours, the animation and the detail may be new but the concept is not. Mercator would not have found these images that revolutionary!

A critique of the GIS view of the world

Just as GIS have grown, so too has a critique of this area of research and development. The critique takes many forms, but the most common is that the real impact of GIS has been to increase the level of surveillance of the population by those who already possess power and control. The claim is that these systems have helped to create a less human world in which people are not treated and judged by who they are and what they do, but more by where they live. In Britain today, whether you can have a particular medical operation might depend on whether your local health authority has a contract with a hospital to carry out sufficient of these operations. In the hospital this is determined for individual patients by looking up the geographic reference of their postcode using a GIS (to find which is their local health authority). The argument suggests that before this technology arrived it was easier to treat people more equitably. Similar examples can be found all over the world, with perhaps the oldest being the automation of insurance 'red-lining' in the United States using postal Zip codes, and the subsequent world-wide use of such techniques to determine the level of credit ratings and insurance premium payments.

   Another part of this critique of GIS concerns the use of systems for precisely locating land rather than people. Global positioning systems (GPS; see Chapter 3) use satellites to pinpoint, with an accuracy of a few metres, the location of little black box receivers. When this information is combined with digitized maps in a GIS, the precise position of the box can be determined. The combination of GPS and GIS has revolutionized surveying in remote areas, and made the capture of digital information about real-world features, such as a rail network, simple (a GPS box is merely attached to the front of a train which is sent round the network). It has even been claimed that GPS systems have saved the lives of shipwrecked people who have been able to radio their precise location to rescuers. So where is the harm? The harm, it is claimed, is that it is just as easy to attach GPS technology to a missile as it is to a child's life-jacket – and there are far more systems on missiles than on life-jackets.

   The critics of GIS suggest that the system's Achilles heel is its widespread use in the military, in particular the way in which these systems could be used to make the process of killing people more efficient. From systems which instantly calculate the line-of-sight of tanks (so that soldiers know where they cannot be seen and who they can see) to the software which guides cruise missiles, the aim is to make somebody else's death more likely. Like almost any other technology, these systems can be used in war. What makes GIS particularly lethal, at least in the imagination of many critics, is its ancestry in the map and those ancient cartographic links with warfare and the demarcation of territory (see Chapter 5). There is a mystique to the power of these systems, a mystique which is fuelled by a lack of information. The use of GIS in the military is secret, so people who know what is happening are not allowed to write about it. Inevitably this engenders speculation.

Secrecy, both in the military and also in many of the commercial uses of GIS, is a key reason why so much criticism has emerged. If people will not tell you what they are doing you may well suspect the worst. In practice, as indicated in Chapter 5, it could easily be the case that many military GIS systems do not work in warfare (GIS are often unstable under pressure) and so the use of arms money to purchase and develop these systems could be saving lives by curtailing the supply of bullets elsewhere! Similarly, commercial GIS may be near to useless at guessing the characteristics of people from their postcode, but it is in the interests of both the GIS industry and its critics to claim that GIS, through geodemographics, can do this. Secrecy hides incompetence as often as it conceals conspiracy, so behind the curtain of high technology there may not be a wizard, but a little man desperately pulling levers to keep the machine working.

Further reading

The 'Bible' for GIS is a two-volume *magnum opus* published, perhaps a little unimaginatively, with the title of *Geographical Information Systems* (Volume 1 of the first edition is subtitled *Principles*; Volume 2, *Applications*) and edited by D. J. Maguire, M. F. Goodchild, and D. W. Rhind (Longman, Harlow, Essex, 1991). In relation to the discussion above, see the chapters by Coppock and Rhind, 'The History of GIS'; Buttenfield and Mackaness, 'Visualization'; Unwin, 'The Academic Setting of GIS'; and Aangeenbrug, 'A Critique of GIS'. For the entertaining origins of the critical debate on GIS, see the series of editorial exchanges in the journal *Environment and Planning A* initiated by P. Taylor (in an editorial first published in *Political Geography Quarterly*, **9** (3) (1990), 211–212). To see how the debate developed read J. Pickles (ed.), *Ground Truth: The Social Implications of Geographic Information Systems* (Guilford, New York, 1995).

The definition of GIS used here is taken from J. Dangermond, 'What is a Geographical Information System?', Chapter 1 in A. I. Johnson, C. B. Pettersson and J. L Fulton (eds), *Geographic Information Systems and Mapping – Practices and Standards* (American Society for Testing and Materials, Philadelphia, 1992). An alternative definition, and in general a good guide to the implications for government, is given by a publication from Her Majesty's Treasury, *An Introduction to Geographic Information Systems* (Central Computer and Telecommunications Agency/HMSO, London, 1994). The implications of

GIS for academia and cartography have been predicted by M. F. Goodchild, 'Stepping Over the Line: Technological Constraints and the New Cartography', *The American Cartographer*, **15** (3) (1988), 311–319; while in the same issue of that journal a more personal view of the history of GIS is given by D. W. Rhind, 'Personality as a Factor in the Development of a New Discipline: The Case of Computer Assisted Cartography', **15** (3), 277–289.

Examples of the use of GIS in military applications are provided by R. A. McLaren and T. J. M. Kennie, 'Visualization of Digital Terrain Models: Techniques and Applications', in J. F. Raper (ed.), *Three Dimensional Applications in Geographical Information Systems* (Taylor and Francis, London, 1989). A good example of the use of GIS in geodemographics is given by S. Openshaw, M. Blake and C. Wymer, 'Using Neurocomputing Methods to Classify Britain's Residential Areas', in P. Fisher (ed.), *Innovations in GIS 2* (Taylor and Francis, London, 1995).

Chapter 8

Alternative views

Introduction

Most maps represent the world in quite conventional ways. Mapping, in this sense, is all about convention. This book reflects this tendency by initially concentrating on traditional cartography and issues such as accuracy, cartographic history, the mapping of territory and the impact of advanced technology on existing practices. In contrast, there are many maps that have not been produced by anonymous government agencies and there are many that have not been drawn using traditional cartographic methods. These maps often represent the world in very different ways from the standard topographic sheet. They may stretch the projection to highlight the areas their authors are interested in: they are not greatly interested in minute details of accuracy and they concentrate on subjects usually avoided in mainstream cartography. Here, examples are taken from ecological mapping and green politics, from human cartography and mapping in Sweden, from alternative atlases of the state of the world, of wars, of the global situation of women, and of the social and economic geography of individual countries. To illustrate how evocative the images of the world created with some of these alternative views can be, the chapter finishes with two examples of a subject rarely discussed and hardly ever mapped: the cartography of genocide and the Holocaust. However, we begin at the opposite extreme, with mapping that is widely discussed and overtly presented to the map-using public: propaganda cartography.

Map propaganda

When is a map an alternative view and when is it propaganda? How 'correct' are traditional views and when is a map not misleading? These are not easy questions to answer or simple words to define. Perhaps the most straightforward reply is that all maps present alternative views and all represent some form of propaganda (meaning to propagate a particular doctrine, usually with some subtlety and often by the state). In practice, however, some maps stand out much more than others as unusual, unconventional or controversial. This is done either by the way in which they present information or by what information they choose to present. Most maps that are unconventional are blatantly so, which is what is meant here by 'alternative views'. Some are not so explicit in their intent: they masquerade as conventional maps, and these can be defined as 'map propaganda'. There is no hard and fast line between these types of mapping and there is huge variation within these typologies. The only way to get a feel for the definitions is to consider some examples and decide for yourself what is different and what is deceptive. The rest of this chapter describes a few such examples and explains a little of where they came from and the motivation behind the production of some alternative views of the world.

The best introduction to map propaganda is a book that tells you how to produce it. Mark Monmonier's (1991) *How to Lie with Maps* does just that. In fact, the purpose of the book is to develop the reader's scepticism rather than to teach dishonesty, but it is more fun to read it as a textbook in deception! For Monmonier, propaganda maps begin with national maps and national atlases, which are usually produced to assert a nation's right to exist and to legitimize its geography. The creation of many new states after the Second World War and in the following era of colonial independence resulted in a spate of national atlases. National atlases typically contain maps of the 'natural' physical geography of the country closely followed by diagrams of its political, economic and social make-up which often serve to present an overall picture suggesting that this country itself is natural (see Chapter 5). Some of the clearest examples of map propaganda come from territorial claims where two countries' maps include the same piece of land within their borders and choose to name it differently: the Falkland/Malvinas Islands is a well-known example, but there are many others, particularly following the end of the communist era in eastern Europe and the re-birth of nationalist identity in its plethora of separate states.

More subtle examples of map propaganda include the use of particular projections to make areas with possible aggressive intent appear huge. The Mercator projection was never designed to show the northern coast of Russia (which was not navigable in the sixteenth century), but Russia (and even more so its previous incarnation as the USSR) can look very threatening when its exaggerated land area is boldly shaded on a world map using that projection. Arno Peters used the weaknesses of the Mercator projection to popularize an equal-area world projection that he had worked on. Among other groups and agencies, the idea of the projection was sold to UNESCO (the United Nations Educational, Scientific and Cultural Organization) (see Chapter 2). This projection could also be seen as propaganda, however, for as Monmonier (1991) and Chapter 2 point out, an equal-population projection would have been a more equitable world map. That having been mentioned, the equal-area projection does highlight the relatively low populations of Africa, Latin America and the Middle East, from which a disproportionate number of UNESCO diplomats come!

Some of the best examples of map propaganda were generated by the Nazi regime in Germany around the time of the Second World War. In 1939 and 1940, maps were produced for the (then neutral) American public showing how Germany was surrounded by aggressive countries, had too little land to exist in and was tiny in area compared to the British Empire. These maps were produced to try to encourage Americans not to enter the war. A very different picture of wartime Germany is given below, of the mapping of the Jewish Holocaust (which the USA's entry into the war did not prevent). If Germany had won the war we may now have been using examples of mid-twentieth-century wartime British or American maps as illustrations of propaganda. English language geography textbooks from the 1920s and 1930s contain many classic examples of propaganda. One of the most famous is Huntington's map of the 'Distribution of Civilization' based, classically, 'on the opinion of fifty experts in many countries' (Huntington, 1924, p. 352). An astute reader can guess what the map tries to show: which countries have 'civilization' and which do not. Huntington's deterministic geography led him to propose prevailing climate as the primary arbiter of 'level of civilization', and subsequently into overtly racist explanations of patterns of distribution on the earth's surface.

How to Lie with Maps also includes (what its author sees as) positive examples of maps being used by environmental pressure groups and social activists. The site of a proposed polluting waste incinerator is plotted on one map with bold concentric circles radiating from it. The use of differently scaled text to mark the radiating circles produces an oppressive image of danger concentrating in the centre of the map. The example from social activism is of a map of local districts in San Diego, California, on which each district bears the name of the country with the closest infant mortality rate to its own. Thus, the outer suburbs are given names such as Sweden, Switzerland and Australia, whereas the inner city districts are labelled USSR, Jamaica and Hungary. This method provides a very dramatic way of getting a simple message across, i.e. that within one city in the United States of America social divisions are as great as are found right across the world. As Monmonier says of these examples, they 'both demonstrate that cartographic propaganda can be an effective intellectual weapon against an unresponsive, biased, or corrupt bureaucracy' (Monmonier, 1991). One mapper's bias is another mapper's truth. Maps such as these can be thought of as 'resistance' maps, rather than propaganda, because their message is more blatant. One rapidly growing form of resistance mapping concerns the environment.

Ecomapping

With the rise of the Green movement over recent decades, the number of researchers engaged in mapping the environment has grown at an incredible rate. Green activists often stress the importance of the mapping of local areas by local people. Behind this viewpoint, the theories of ecological mapping provide very different ways of looking at the world. These also tend to be both conspiratorial and romantic in nature. A pioneer of ecomapping is Doug Aberley:

> if you were entirely cynical, you could view the appropriation of mapping from common understanding as just another police action designed to assist the process of homogenizing 5,000 human cultures into one malleable and docile market. As a collective entity we have lost our languages, have forgotten our songs and legends, and now cannot even conceive of the space that makes up that most fundamental aspect of life – home. (Aberley, 1993, p. 2)

Aberley argues that maps can be used to make starkly visible the harm that 'big business' and 'centralized government' do to the environment and to social

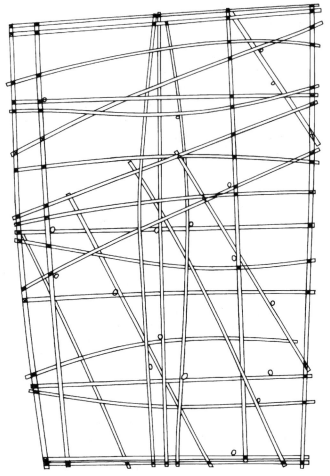

Figure 8.1   Marshall Island sea chart of sticks and shells (redrawn from Aberley, 1993, p. 10).

justice. Maps, he says, can magically communicate the larger evils of a society divided by poverty, race, sex and age. Maps can also be used to stir up action by presenting Utopian visions of how the world could be. He argues that anyone can map and that everyone should, while a great deal less reverence should be paid to professional and scientific cartography. What matters to Aberley is that people start drawing their own maps of their 'own' neighbourhoods and begin to understand the 'bioregion' they live in. Aboriginal mapping is used as the inspiration for this understanding (Figure 8.1).

Aboriginal mapping is largely ignored in conventional cartographic history. Perhaps this is because the maps are so unfamiliar, being constructed with material such as string, blood and stones more often than paper and ink. Upon closer inspection, however, these maps often turn out to be intricate descriptions of local environments. Polynesian (Micronesian) maps show the positions of stars that could guide their sailors between islands in the Pacific. Many Eskimo

Box 8.1   Personality box – Doug Aberley: map-maker and bio-regionalist

Doug Aberley's book contains many more instances of aboriginal mapping than those considered here. The book is a mapping primer for people interested in green politics who also want to change the world. It contains a collection of articles which provide examples of people mapping their own environments. These range from individuals in cities or villages, to pressure groups using geographical information systems to argue a point, or native Americans trying to reclaim their territorial birthright. At the foundation of Aberley's work is a profound dislike and distrust for conventional mapping and the consumer society. This is best explained in his own words:

> When the fundamental importance of perceiving real and imagined space is compared to what passes for most mapping today, a huge separation is apparent. In our consumer society, mapping has become an activity primarily reserved for those in power, used to delineate the 'property' of nation states and multi-national companies. The making of maps has become dominated by specialists who wield satellites and other complex machinery. The result is that although we have great access to maps, we have also lost the ability ourselves to conceptualise, make and use images of place – skills which our ancestors honed over thousands of years. In return for this surrendered knowledge, maps have been appropriated for uses which are more and more sinister. Spewed forth from digital abstraction, they guide the incessant development juggernaut. They divide whole local, regional and continental environments into the absurdity of squared efficiency. They aid in attaching legitimacy to a reductionist control that strips contact with the web of life from the experience of place'. (Aberley, 1993, p. 1)

Doug Aberley is also not too enamoured of academics, often describing them as 'eggheads' or 'isolated scientists'. His review of current cartography books contains several amusing criticisms of the ability of university lecturers to write clearly! To understand what Aberley is doing, it is important to know where he is coming from. He worked for many years as a town planner in Hazelton on the Skeena River in the far north of British Columbia, Canada. He is also an active environmentalist (he would say bioregionalist) and map-maker. Just as Aberley stresses the importance of knowing about your local area, it is important to know where other people come from, in order to put what they say in context. It is doubtful that a book like this would have been written anywhere other than on the western coast of North America.

(Inuit) maps were scaled to show travel time on foot. Some tribes from Siberia (such as the Tchuktchi) used maps, not orientated north or south, but according to the flow of the dominant river in the area.

Ecomapping combines such aboriginal techniques with input from more modern trends such as the Garden City Movement which produced many maps at the turn of this century of how cities could be. These alternative maps combine age-old interests, such as the location of water sources, with new concerns, such as the distribution of pollutants, to produce images that are designed to wake people up to action. Maps have always presented pictures of 'truth' and just as many people have many different truths, so there are many maps to be drawn.

A further example of this kind of resistance mapping is that promoted by the Common Ground organization which was established in London in 1983 to promote a common cultural heritage of areas in Britain, celebrating their local distinctiveness and their links with the past. Their motto is 'know your place – make a map of it!' and they have produced a booklet describing just how this can be done (Greeves, 1987). Again, the maps that this project promotes will serve particular purposes, but a purpose different from official maps of areas. One purpose, which is made explicit in the booklet, is to use the maps as a tool to try to stop planners, far removed from the locality, allowing changes to be made to parts of the country. In practice, they have been used most often to preserve rural parts of southern England from the perceived threats of development (Figure 8.2). In many cases there is a grey line between resistance mapping and propaganda.

Humanist cartography

What kind of alternative mapping people propose very often depends on where they have come from. Here we discuss an area of alternative mapping that developed out of concerns with social policy and academic cartography. *Human Cartography* is the title of a book written by Jan Szegö who worked for the Department of Planning and Natural Resources of the National Board of Housing in Sweden. 'Human Cartography' was the term Szegö (1987) gave to mapping where the focus was on people, where they lived, where they were going, what they did. Conventional cartography has been focused on land, even if there has been a human aspect in the mapping of land ownership or the navigability of the terrain for armies of men. Humanist cartography, in contrast, should concentrate on the human experience of space and portray the (often highly diverse) human encounters with 'reality', rejecting the view that behaviour (and, therefore, features such as population distribution and the location of industrial activity) is governed so totally by the framework of the earth and the 'tyranny of distance'. During the nineteenth century, interest in population statistics grew. As people who had been peasants became consumers, the relative value of land to human life fell, and human geography began to matter. The emergence of detailed census cartography following the Second World War grew out of these shifting priorities. But Szegö argued that much census cartography was still not concentrating enough on people (Figure 8.3).

The basic tenets of human cartography are that the mapping of people should be based primarily on showing movement over both the short and long term. The daily and yearly ebb and flow of people is of vital importance to human lives. The population should be seen as a 'vertically rising stream in space–time' in which each individual's life history is marked out as a path in a three-dimensional structure. The combination of many paths creates a picture, a picture

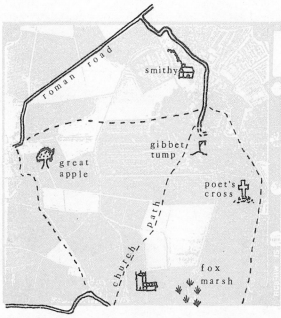

PARISH MAPS

CELEBRATING AND LOOKING AFTER YOUR PLACE

−1−

A
COMMON GROUND
PUBLICATION

Figure 8.2 The Parish Maps brochure (front cover) (redrawn from Greeves 1987).

which human cartography should draw. In arguing for this perspective, human cartography follows a Swedish academic tradition of studying diffusion which is most closely associated with Torsten Hägerstrand, but it does depart from this somewhat, with its greater interest in producing images, rather than models, of these processes.

The book *Human Cartography* gives examples of numerous ways of mapping statistical information about people. The use of areas that encompass equal numbers of people is advocated and a wide variety of symbols are developed to fill these spaces. Lines and arrows are used repeatedly to connect the places where

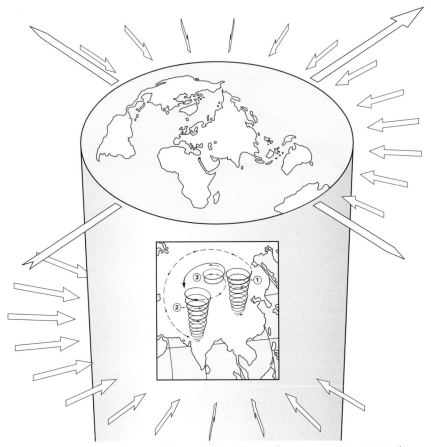

Figure 8.3 Human cartography and a space–time diagram (redrawn from Szegö, 1987).

people live with their workplaces, or to show the average direction of commuting out of an area. The use of maps to show longer historical trends is also suggested and simple diagrams of space–time evolution are included. The book is full of beautifully drawn sketch maps designed to encourage the reader to begin creating their own images and to break away from the usual carto-graphic conventions.

If there were one major criticism of human cartography it would be that these methods work well for showing the space–time development of a small com-munity or the commuting patterns between (and to) a couple of towns, but where an overview of an entire country is wanted – such as could be seen from the back of a fairy tale goose (see Box 8.2) – then much simpler graphical methods have to be used. The space–time history of even just a hundred people presents too complex an interwoven picture to be drawn on a single page of paper, while the detailed patterns of commuting within a major metropolis are very difficult to depict clearly with arrows which suggest stream-like flows (streams tend not to cross one another). In many ways, creating the cartography

Box 8.2   Personality box – Janos Szegö and human cartography

Szegö has written three major books on cartography and each one begins with a short story. His first book was *A Census Atlas of Sweden*, published in 1984, which begins with a famous Swedish fairy story in which a 12-year-old boy looks down upon a patchwork quilt of his country from the back of a goose – a literal bird's eye view. What mattered most for Szegö's theory of human cartography, however, was not the bird's eye view, but the story of the fairy tale, which tells of interdependence between people's (and animals') lives. This is something which is missing from much modern cartography, which often merely tabulates statistics using coloured dots in place of lists of numbers. Many of the maps in his census atlas also reduce people's lives to this form of visual tabulation, but at least in the author's mind, and often to the reader, things are different:

> Sometimes peculiar things happened during the lengthy work with analyses, comparisons, drawing, checking. It sometimes seemed as if the red symbols and yellow ground colour of the urban landscape faded away and were replaced by forests, lakes and fields, criss-crossed by roads, which turned out to be filled with streams of sleepy early-morning travellers in cars, buses and trains, on their way to work in the towns. There were times when the coloured areas on the colour ink-jet map were suddenly obscured by white summer clouds which seemed to scud in from nowhere between the map and the author's eyes, and among which he could glimpse the sparkle of the sea by the coast, the rivers which rolled down to meet it, the towns and villages and people. Sometimes old people materialized out of the map of Norrland and observed with melancholy the exodus of the young towards the coast and the south. From the diagrams which display households suddenly appeared a throng of people who with muted voices told of their lives, of their loneliness, of their joy in their children and of their hopes on their behalf'. (Szegö, 1984, p. 20)

The second book, *Human Cartography*, was published three years later and describes some ways in which a little of this dream could be realized (see main text). Later, in *Mapping Hidden Dimensions of the Urban Scene* (1994) many of these techniques are used, concentrating in particular on the depiction of the changing population of the town of Malmö in southern Sweden. Like *A Census Atlas of Sweden* published ten years earlier, this book also begins with a story of a view from the sky, but this time from a hot-air balloon which contains a magical machine. The machine is a scanner which allows the locations of people to be viewed, through walls and roofs, as coloured dots – just as street lamps or car headlights may be seen at night. At night most dots shown by the scanner are stationary, but a new one suddenly appears when a child is born. Then, with the start of another day, they begin to move frantically, quickly forming lines of dots commuting to work, going to school or milling around when shopping in local neighbourhoods. The dream of human cartography is to show the structure of the patterns and movements of these millions of dots which are people, both daily and over the decades.

of people is a far more complex challenge than producing the cartography of land. Most challenging, perhaps, is the mapping of *all* the peoples of the world. This is represented by another section of the 'broad church' of alternative cartography.

The 'new' world atlases

A key way in which alternative mapping has emerged and has been published widely has been in the form of alternative atlases. Given the controversy that surrounded the introduction of the Peters Projection and Peters' *Atlas of the World* (see Chapter 2) it is surprising how little anger has greeted the arrival of far more alternative atlases. The assumption has to be that they were not taken very seriously in global debates and many of these atlases were not reviewed in the academic cartography journals. This is a pity because this is another area of cartography that is growing very quickly. Soon more students may be directed to look up facts about the world in alternative atlases, than are using the standard school-board approved atlases of the world.

Perhaps the most well-known alternative atlas is the *State of the World Atlas*, first published by Michael Kidron and Ronald Segal in 1981 and quickly followed by a 'new' atlas in 1984. What is most striking about these atlases is that they use area cartograms on many pages to show how inequitably the world's wealth, weapons and food are distributed. Each cartogram has been designed by hand to keep the shape of the world familiar while still showing clearly how unfair the current distribution of resources and power is (see Box 8.3). These are some of the first detailed world maps many children growing up in the 1980s studied at school. Thus the effect of these images may be with us for some time to come (Figure 8.4).

In many ways the State of the World Atlases are not alternative. In essence, they simply show statistics for a couple of hundred sovereign states which are generally accessible from United Nations Yearbooks. What is alternative about these atlases is the subjects they choose to cover and the way in which they present those subjects. Pictograms are often used with small symbols to show, for instance, people, rockets or coins. The atlases received a warm welcome when published, but, as the authors themselves say: 'most flattering of all has been the spate of imitations in books, newspapers and on television, confirming the value of our method, even where views very different from our own have been advanced' (Kidron and Smith, 1995).

After the *State of the World Atlas*, Pluto Press, the publisher pushing this alternative view, moved on to other similar projects (see Box 8.4). Notable among these was the *Women in the World International Atlas*, drawn by Joni Seager and Ann Olson and published in 1986. This atlas used a similar format to its predecessors but covered a very different set of subjects ranging from the geography of women in the military to the world map of female circumcision. In terms of the cartography, the atlas differed from its predecessors in that around the edge of each map there tended to be far more adornment. Graphs, tables and pictograms surround each map. Sometimes they even took over whole pages, showing

Box 8.3 Contemporary mapping box – Cartograms: changing the shape of the world

Area cartograms, maps which are 'distorted' to use area explicitly to show some information about the relative values at places, have a long history. Many traditional maps are 'equal land area cartograms', but they are often not thought of as such. Most cartograms, however, scale the map areas to be proportional to population, so that statistics can be shown on a projection which does not discriminate against people living in densely populated areas. The easiest way to understand how cartograms work is to draw one yourself.

(continued)

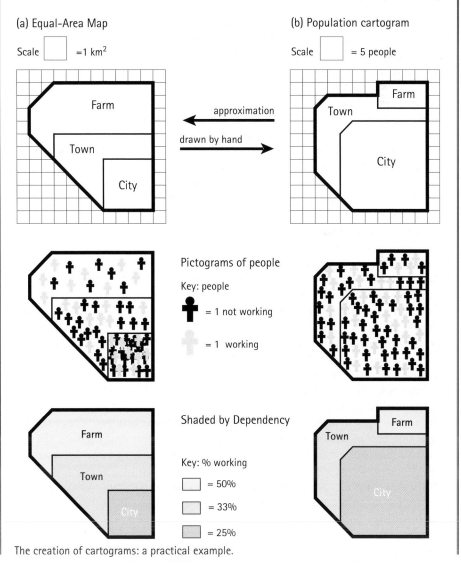

The creation of cartograms: a practical example.

(continued)

The figure presents some population statistics and a map of a fictional island drawn on squared paper. Its area is 20 km² and 100 people live on the island. For administrative purposes the island is divided into three districts called Farm, Town and City. The mean population density of the island is therefore 5 people/km², but from the statistics given it is evident that most people on the island live in the City at a density of 15 people/km². Thus, if we use a traditional equal land area map of the island to show social statistics we will be visually under-representing the bulk of the population who live in its City.

To create an equal-population cartogram of the island by hand it is best to start with a simplified equal-area map drawn on a grid, as shown in the figure, and then to try to replicate this image as closely as possible while still scaling districts to their correct areas. The equal-population cartogram of the island shown here was drawn by hand, starting with the City as the simplest shape. Note that the Town is then added in such a way as to still separate the Farm from the City.

The effect of using the cartogram instead of the maps is then illustrated by two examples in the figure. The first shows two 'pictograms' of the distribution of the population which differentiate between people in and out of work. On the traditional map it looks as if roughly the same number are in each group, whereas on the cartogram the large non-working population of the city is given enough space to be seen. The second example shows the same information but in the form of a choropleth map and cartogram.

the most complex of statistics such as the proportions of women working in four levels of education or as nurses, doctors and gynaecologists for dozens of countries.

In one sense these 'alternative' maps can also be seen to have strong historical precedents. Just as ecological mapping traces its history to aboriginal maps and the introductions to human cartography begin with fairy tale images, there are precedents in medieval mapping for the 'new' state of the world atlases and their successors. Medieval mappers were often unconcerned with equal-area projections. Their maps scaled and orientated areas by their social importance (e.g. the *mappae mundi* centred on Jerusalem (Figure 1.4) and the map of the world (Figure 1.2), which ignored virtually all known territory outside the Roman Empire). Some regional medieval maps also tended to contain a great deal of extra material around their borders. These images were not extraneous to the maps, but contained more information (in the form of sketches of the local inhabitants, coats of arms of the local gentry and pictures of the primary craft industries, for example) about the social, political, economic and ecological priorities of their times. Maps may be alternative, but they are rarely entirely new. One of the oldest inspirations for map-making, and now a subject of increasing interest for mapping itself, is warfare.

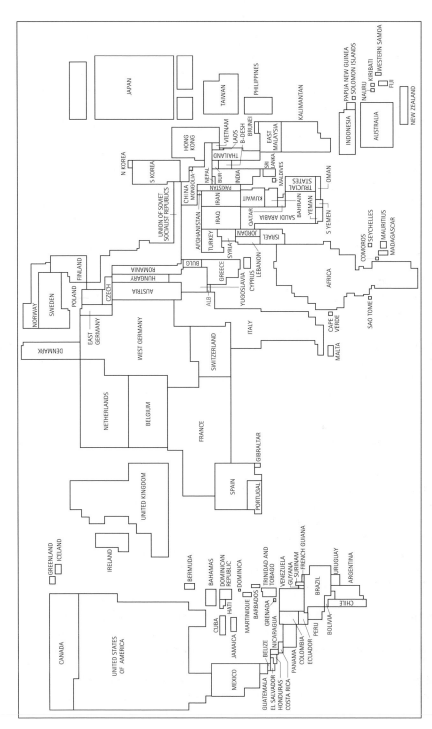

Figure 8.4 The distribution of world trade (redrawn from Map 18 of Kidron and Segal, 1984: Trade Power).

Box 8.4   Personality box – Michael Kidron and the Pluto Press Project

Michael Kidron published the first *State of the World Atlas* with Ronald Segal in 1981. The work for the atlas was coordinated by Pluto Press and published by Pan Books. In their introduction to this atlas the authors explain where the ideas had come from:

> The *State of the World Atlas* stems from the rich tradition of political atlases, born out of war or the threat of war, and out of the widespread interest that this excited in military strategy and deployment. In time the tradition came to embrace economic, social and cultural factors; reflecting and promoting the diffusion of knowledge about the world at large which has accompanied the spread of mass education, new forms of communication and international travel' (Kidron and Segal, 1981, preface).

Michael Kidron was general editor of Pluto Press, under the auspices of which a large number of the current series of political atlases have been published. These have ranged from the international, covering topics such as war or the inequality between the sexes, to national atlases such as the *State of the Nation Atlas of Britain* (see main text). This format of publishing has proved very popular and has been widely copied in both format and sometimes even title. *The State of the USA Atlas* written by Doug Henwood and published by Penguin is a clear example of the spread of the style of Pluto Press. The fifth revised edition of *The State of the World Atlas* was published in 1995 by Penguin, over a decade after the original book had been released. There is certainly a market for such alternative atlases.

The cartography of war

There is a connection between cartography and war other than mapping for the purposes of military activity: maps about war. Included in this category could be historical atlases showing the campaigns of armies and the overthrow of nations. The *Times Atlas of World History* is the most well known among a plethora of examples of this genre. In this section, however, the interest is in mapping which does not merely document war but which actively opposes it or uncovers its horrors. Just four examples are given here but many more exist. The first is another Pluto Press project – *The War Atlas*, drawn by Michael Kidron and Dan Smith (1983)– and the second is a very different book – *The Strategic Atlas of World Geopolitics*, written by Gérard Chaliand and Jean-Pierre Rageau (1986).

The War Atlas was published in 1983, at the height of the nuclear arms race. In cartographic form it documented the history of wars up to the geography of the 300 wars that took place in the world in the four decades from 1945. From the maps it is apparent that very few countries have been at peace over even this short period. Most importantly, however, the atlas mapped the 40 000 strategic sites throughout the world identified as nuclear targets by the USA (places the Americans planned to strike and places they thought would be others' targets).

From just this one map it is clear that very few people would escape the instant effect of a nuclear war. However, the atlas goes on to show where poison gas and nerve agents have been used in wars across the world; how military satellites track the earth's surface; the size of different armed forces in each country; the locations of 'military advisors'; the geography of military spending and of the arms trade. Cartography can produce emotive images – images which make a political point. However, this is not simple propaganda because the aim of these images is explicit: to show just how much human effort goes into making war possible and the scale of human misery that results.

A few years after the publication of *The War Atlas*, what was claimed to be the world's first strategic atlas was published, *The Strategic Atlas of World Geopolitics*, subtitled 'world geopolitics: a new and exciting survey of the polit-ical realities of today's world' (Chaliand and Rageau, 1986). Although not strictly a pacifist atlas, it did highlight the dangers of war and document just how many armed conflicts there have been. One of the distinguishing features of this atlas was that most of the maps used an azimuthal projection centred on the pole (Figure 8.5). This projection highlights the close proximity of the United States to the former Soviet Union and also shows the trajectories of the projected flight paths of many intercontinental ballistic missiles as straight lines across the northern polar regions. The atlas not only maps the entire globe using a variety of projections but also includes many inset maps of small areas of great strategic importance, even including a street map of Beirut, which at the time of publica-tion was a city virtually destroyed by conflict.

The ultimate mapping of war is to be seen in the mapping of holocausts. Literally, holocaust means the wholesale destruction of a people, especially by fire. Two examples of holocaust mapping are now considered. The first is *The Atlas of the Holocaust* written by Martin Gilbert and published in 1982. This atlas tells the story of the genocide of the Jews in Europe between 1933 and 1945 and is one of the most evocative atlases ever produced. The second example is William Bunge's *The Nuclear War Atlas*, telling of the holocaust to come and first published in 1988.

At one level *The Atlas of the Holocaust* presents, in exhaustive detail, graphical depictions of the evidence of the genocide, largely of Jews, which took place in Europe between 1933 and 1945. At another level the atlas goes beyond this by trying to begin to show graphically what the sheer numbers of deaths meant. Previously this had largely been done by physical memorials and museums of the Holocaust and through media such as film. One aspect of the Holocaust which these filters had difficulty conveying was the level of geographical organization involved in its execution. This is illustrated in Figure 8.6 which shows two maps redrawn from the atlas. The first shows the movements of one family, the Hirschsprungs. The mother of the family, Helene, was born in the nondescript village of Auschwitz in 1909 and married a man from Cracow in what was then the Austro-Hungarian Empire. They emigrated to Amsterdam in Holland where their children were born in the 1930s and then fled with their children to Lille in France when the war broke out. On 15 September 1942, along with another 1000 Jews, they were seized by the SS and transferred to Malines before being deported to

Figure 8.5 The polar azimuthal projection (redrawn from Chaliand and Rageau, 1986, p. 11).

Auschwitz to be killed. The second maps shows all the deportations to Auschwitz in a two-week period a month after the Hirschsprung family was seized.

By showing a map of the path to death of one family and then of the thousands of families who were systematically transported in a short space of time, *The Atlas of the Holocaust* attempts many things. It tries to show us the story of individual tragedy and of how that could be related to the movements of many thousands of families and individuals without losing all of the meaning. It tries to show what level of organization was required to arrange the deaths of so many millions of people in such an apparently orderly fashion. Most importantly, perhaps, it presents the evidence of an event – stretched over hundreds of maps of Europe – that could not have gone unnoticed by the people living in these places. It makes us question the received history of the Holocaust by putting the event in our faces – on the map.

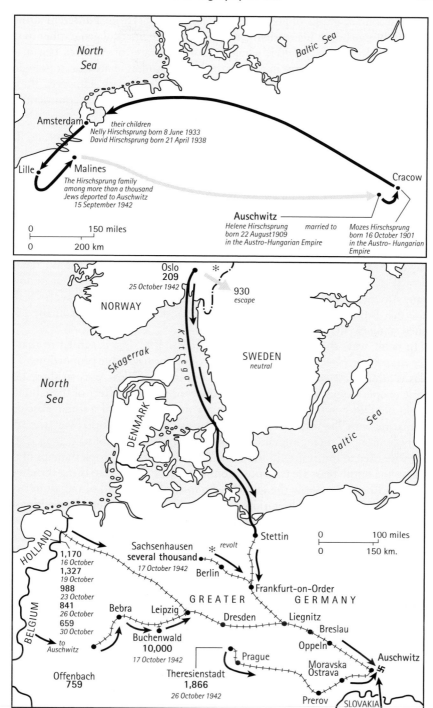

Figure 8.6   Geography and the Holocaust (redrawn from Gilbert, 1982).

The second example, of possible future holocausts, is from a different author. William Bunge began his academic studies as a mathematical geographer in the late 1950s. In February 1965, when the United States of America began the bombing of North Vietnam in earnest, he left his studies of abstract topological space in the classroom and moved, full time, into peace work. However, Bunge never quite left mathematics completely behind. For instance, he comments on the American protests against the Vietnam war:

> the peace chant against President Lyndon Baines Johnston was instantaneous, if at first just a tiny minority voice; 'Hey, hey, LBJ, how many kids did you kill today?' I have recently computed the answer to that old question. On average we Americans killed 300 Vietnamese children each day, for ten years reaching a grand total of 1,000,000 and they did not kill one of ours. (Bunge, 1988, p. xviii)

During the Detroit riots of July 1967, Bunge began work on a 'peace book', *Fitzgerald: Geography of a Revolution*, about the Detroit neighbourhood in which he was living. A year later he was listed by the United States Senate Anti-Subversive Committee as one of 65 people (nationally) who should not be allowed to speak on university campuses. This was one of the highest honours many American academics of that time felt they could achieve. It was also at this time that Bunge began work on *The Nuclear War Atlas* which was finally published some twenty years later.

In many ways the central aim of *The Nuclear War Atlas* is similar to that of *The Atlas of the Holocaust*: to convey a horror of unimaginable proportions. The atlas is attempting to envisage the sheer gravity of nuclear war; to show how many people would be killed and just how enormous the effect of these bombs would be. Dot graphs are used to compare the fire power of a single Poseidon submarine with the total fire-power used in the Second World War (the submarine contained three times more explosives in equivalent tonnage than *all* the explosive used in munitions in the Second World War). Choropleth maps of the Hiroshima bomb blast are transposed onto a map of Chicago to bring the effects 'home'. Other maps show the global geography of the spread of radiation from nuclear 'tests' and the proliferation of nuclear weapons to other states. This atlas is an excellent introduction to the understanding of alternative mapping and the efforts of one academic to produce a different image. Although alternative maps might today appear commonplace, the right – both intellectually and politically – to produce another view was hard fought for.

Summary

Maps provide powerful images. For people who want to change the way we think about the world, changing our map of the world is often a necessary first step. For those defending the status quo, maps are equally important. Often very subtle touches show just who is in power. British maps of South Africa before the Boer War were printed in two colours: red for the roads and blue for the rivers, upon a white background. The red, white and blue images literally stamped Britain's authority to rule over that colony. After that war this subtly continued with maps of the Republic of South Africa under apartheid using a small typeface to label large black townships while small white settlements with considerably fewer residents were annotated in a larger bold type. Today the cartographers of South Africa have a new problem. With eleven official languages, 'politically correct' maps of South Africa may be too cluttered with text to be usable.

True alternative mapping is rarely produced by the official cartographic offices of the state. More often it is single individuals who are inspired to draw a different picture of part of the planet, but these individual efforts can be grouped. Here mapping from ecological, humanist, socialist and pacifist cartographics have been grouped along with examples of feminist, aboriginal and fascist cartography. There are a large number of ways of labelling human views of the world and for each label alternative ways of mapping have evolved. However, this list is far from exhaustive and new views are being drawn all the time. Although most alternative maps can be neatly pigeon-holed with their predecessors, a few are very different. These are the new maps which really make others stop and think.

Further reading

M. Monmonier's *How to Lie with Maps* was published by University of Chicago Press, Chicago, in 1991. A useful introduction to environmental mapping is D. Aberley's *Boundaries of Home: Mapping for Local Empowerment* (New Society Publishers, Gabriola, British Columbia 1993). J. Szegö's three books on alternative census cartography are well worth reading: *A Census Atlas of Sweden*, *Human Cartography*, and *Mapping Hidden Dimensions of the Urban Scene* (all published by the Swedish Council for Building Research, Stockholm, 1984, 1987 and 1994 respectively).

The State of the World Atlas was produced by M. Kidron and R. Segal and published by Pan Books (Heinemann) in London in 1981. *The New State of the World Atlas* was written by the same authors and published by Pluto Press in London in 1984. A year later, *The State of the Nation: An Atlas of Britain in the Eighties* was produced by S. Fothergill and J. Vincent (Pluto Press, London, 1985). Next, J. Seager and A. Olson wrote *Women in the World: An International Atlas*, also a Pluto Press project (Pan Books, London, 1986). M. Kidron and D. Smith published the first edition of *The War Atlas: Armed Conflict – Armed Peace* in 1983. Their *New State of War and Peace: An International Atlas* was published by Grafton Books, London, in 1991 and, in 1995, Penguin (London) published the fifth revised edition of *The State of the World Atlas*.

The Nuclear War Atlas was written by W. Bunge (Blackwell, Oxford, 1988) while *The Strategic Atlas of World Geopolitics* was written by G. Chaliand and J.-P. Rageau (Penguin, London, 1986). *The State of the USA Atlas* was written by D. Henwood (Penguin, New York, 1994) and subtitled *The Changing Face of American Life in Maps and Graphics*. Finally, and definitely essential reading, *The Atlas of the Holocaust* is by M. Gilbert (Michael Joseph, London, 1982).

Chapter 9

Representing the future and the future of representation

Introduction

The lessening of institutional activity, and the decline of 'corporate' mapping indicated in the last chapter, and also alluded to in Chapter 5, is the most important recent development in cartography. Increasingly, maps are being drawn by individuals for themselves rather than by companies and government organizations. For virtually the whole of cartographic history (up to now), the opposite balance has been in force, with mapping generally being undertaken by initiated craft guild members, by employees of wealthy patrons and by government agencies. Although individuals may have had their own 'world-views', the representation of such interpretations of reality in map form was an expensive luxury. As a result the prevalent 'correct' view disseminated within almost all cultures and throughout most societies' histories up to the present day, has been that of the powerful and wealthy elite who could afford the rendering of their conceptions in map form.

The recent turnaround has been partly due to the rise in the numbers of people needing, and also willing and able, to create their own map products. Such people are now largely working from home on leisure time activities or are engaged in research and development in small teams without the resources to contract a professional cartographer. It has been the boom in the desk-top publishing industry that has allowed these people to undertake technically sophisticated production away from the traditional map-making workplace and without using the traditional expensive tools and processes. People are drawing maps more and more for their own personal use. Maps, as a means of portraying and conveying information have become more common in popular magazines, in newspaper advertisements, on television, in children's essays (often now done for homework or student course-work), in advertising brochures, on hoardings – even on the paper place-mats in fast-food restaurants.

Thus, a new renaissance in mapping is apparent. New technology has made it easier for many people to produce professional-looking maps, often using 'freehand' type computer packages (considered below) rather than the more complex geographical information systems discussed in Chapter 7. Furthermore, there has arisen an increasing confidence in using and producing graphics generally, by the first generation to grow up in a society where the dissemination of information has been so dominated by television: a generation whose offspring, in most of the western world, are now being introduced to an even wider range of electronic imagery and image handling. In the past, the majority of human communication beyond small groups was achieved through reading and writing. With the advent of broadcast media, the receiving side of the 'communication flowline' (which involves the broadcaster – the initiator of the message – and the sending of a message to a passive recipient), changed to a listening, and later viewing, mode of data acquisition, as opposed to the traditional reading mode. Current developments in information technology and knowledge transfer encourage drawing over writing. This is the context within which we must consider the future of maps and other such visual representations of the world.

New tools and new data

The technical side of the late twentieth-century renaissance in cartography has been well documented (see Chapters 3 and 7). The consequent 'democratization' of cartography has taken place as tools for map production become more widely available. Such tools include the widely used drawing packages such as Adobe Illustrator and CorelDRAW available on Macintosh and PC-based microcomputers; cheap computer-aided design/drafting (CAD) software such as AutoCAD and Microstation; and versatile viewing packages such as VistaPro, which allow for the aesthetic rendering of terrain data. To these can be added the extension of graphics handling ability within word processing, spreadsheets and database programs which can allow for easy creation and incorporation of maps into text- and number-based documents. The all-round ability to manipulate, save and promote existing maps presented by a variety of products from CD-ROM interactive atlases to HTML (Hyper-Text Markup Language) and VRML (Virtual Reality Modelling Language) coded pages on the Internet is further evidence of the unconscious acceptance of cartographic products into the general areas of digital data handling. These tools are, of course, still restricted to particular, privileged sets of people living in certain parts of the world, but their proliferation is rapid and can be expected to accelerate in the future as the reproduction costs of the new technology are, theoretically, very low. Whether access to the data required for such mapping also proliferates at the same rate, and whether access to the education and the time required to participate in this digital revolution also become more widespread, is, perhaps, more doubtful.

These new tools are also extending the range of products used to represent world-views. Our own direct experience of the world can be seen as four-dimensional in nature, considering planimetric position, height or depth, and a time factor (see Chapter 6). The development of cartography into areas such as terrain modelling, which tries to take account of the relief of the earth's surface in its depiction of a picture of reality, and dynamic mapping, which uses animation techniques to show time sequences inherent in the data, has led to calls for a reassessment of cartographic theory and practice, so that the collective wisdom of centuries of map-making expertise can be successfully extended to meet the demands and opportunities of new technology. It remains to be seen whether cartographers as a professional group are capable of responding to the need for a re-evaluation of their craft or whether the increasing numbers of 'do-it-yourself' map-makers will pass them by without a second glance. Neither those who used to control much of what was mapped, nor many of those who have had little say in recent history, may benefit from the recent technological changes. It is also debatable whether the peoples of less prosperous areas, for example, will benefit from the information technology revolution in general, and the cartographic revolution in particular. But cartography is changing theoretically as well as technologically, and further possibilities with even wider implications arise from this.

The four-dimensional world-view is inherent in the development of the new, more human, cartography described in Chapter 8 as well as in the world of the flight simulator. For instance, the lives of people can be visualized as threads running through time, although to depict this using a static map on a computer screen the height of the surface of the earth may have to be ignored. If this is done then the thread of a person's life traces a path in three dimensions (the third dimension of relief being replaced by one depicting time), beginning at their birthplace and ending at the location of their death. The same computer software that is used in physical geography to depict the pattern of boreholes through a complex geology or the intertwining of currents in a river can be used to visualize the movements of a large family through many generations or the diffusion of particular ideas and fashions among the populations of settlements over time. Again, it is questionable whether practising cartographers will be able to accept the splintering of traditional approaches into the myriad of directions, exemplified by such space/time-mapping, which are currently being taken. The danger in not accepting this and ignoring such new developments is that much common practice may have to be re-discovered by the 'do-it-yourself' map-makers who may initially reject, ignore or simply not be aware of past practices. The future of representation will depend as much on teaching and learning as on new technology and the changing nature of societies in the world. However, to understand the latter we must first realize what barriers exist to the proliferation of new technology beyond its initial cost.

Of equal importance in cartographic development to the proliferation of software and hardware, therefore, has been the expansion in the availability of digital spatial data outside of 'official' walls. As more spatial data become available, more maps can be made. These data come from government (national and local); from private companies (including some long-established map producers such as Bartholomew in the UK and new companies created solely as digital spatial data retailers such as GeoSystems in the USA); from remote sensing of all types; from data collection exercises by individual students and by large-scale research projects; from environmental monitoring; and through 'freedom of information' legislation (which is more apparent in the USA than in Europe). Although these data do encompass topographic and general reference-type material, it is significant to note that very little detailed and comparative digital information on human geography is available on a world-wide basis.

The majority of maps today are produced by the people who will eventually use them: the notion of specialist agencies and institutions with the expertise, authority and access to spatial data all denied to the general public is being replaced with a realization that the tools, data, perceived needs and individual-istic world-view, all prerequisites to any mapping project, are now under the control of a vast number of citizens in the developed world. The humanist view described in the Introduction to this book confirms that there is less reliance on a 'scientific' viewpoint which purports to give a 'truthful' picture of the world (such as portrayed on maps produced by national mapping agencies) and more acceptance of the individual view of the world as represented on more ephemeral, less institutionalized maps. In fact, many academics would regard the official mapping as more 'incorrect' than the personal maps.

New roles and new maps

This wide and disparate set of map-making and map using activities echoes the fact that this is a time of more diverse and extensive geographical thought – influenced by more pluralistic views in politics, economics, sociology and the many other disciplines affecting geography. Despite the philosophical changes to such subjects and to geography, however, maps are still seen to be of central importance. The most recently published Open University geography textbook series begins with a discussion on making maps as part of an example of the attempts of indigenous people to gain political power over their lives (see Further reading). An imagined scene in a company boardroom in Honduras is described in the book to introduce new students of geography to the power of maps as representations of the world:

> A group of company executives stands around a table and peers down at the map laid out before them. This act in itself may make the place seem very vulnerable, to their eyes and to their plans. But what makes the situation even worse, from the point of view of the indigenous groups, is that they are not even on that map! It is easy to imagine how much simpler it must be to plan – to cut down the forest here, to open up a plantation there, build a road across this way – if there is no trace, as the point-ers are wielded and the lines are drawn, of the density of settlement and use of this place by those who are already living there. So the new map is designed to make others see, and understand, the place differently.
>
> A more general lesson we can take away from this episode is that maps are a means of representation and every individual map embodies a particular way of under-standing, a particular interpretation of the place it is depicting. What this means is that maps are 'social products' (Massey, 1995, p. 20).

The new map referred to above is a map that was drawn by the indigenous people of the region and which was peppered with the names of their commu-nities and areas of subsistence, so that space on the paper no longer looked blank nor appeared so easily available for development. It is worth noting that, just at the time when technical changes have made the democratization of map pro-duction possible, students on possibly the largest university geography course in the western world (at the Open University) are being taught that maps are essentially social products. 'Technically correct' maps are usually drawn in universities, by governments, or by large companies and these are often the maps of authority. However, this interpretation does not incorporate recent changes which means that maps which look technically correct (even if, in a traditional senses, they are not) can now be drawn outside of such organizations. Maps are both social and technical products, just as cartography can be seen as both an art and a science.

Cartography in multi- and hyper-media and in virtual reality

The impacts of technology on cartography have come from numerous directions. Although primarily engaged in replicating the manual techniques of map-making, it is clear that the opportunities for developing new working practices afforded by the type of computer package mentioned on p. 157 are extensive. The expansion of possible interaction with spatial data beyond the senses of vision is already afforded by the use of multi-media tools. A number of CD-ROM and on-line map products exist which can, using 'hot links' and database navigation techniques, connect maps with images, text, animated sequences, video clips and sound tracks. Using the current abilities of computer graphics hardware and software as exemplified in the entertainments industry and in the military, it is easy to envisage methods of accessing and using spatial data in an immersive environment where reconstructed or fantasy landscapes are beamed to all human senses – vision, hearing, balance, touch and smell. Virtual reality is the name given to such technology which uses headsets, tactile gloves, earphones and, occasionally (e.g. in amusement parks), motion-inducing seating. Cartographers have a potential role to play in both supplying data for such simulations and in advising on methods of presentation of the data: but, as with contemporary digital mapping, the opportunities for the users themselves to create, present, navigate and interact with spatial data are extensive and are becoming the standard mode of virtual-reality use.

For some time in the publishing world there has been considerable speculation that the paper-based book is nearing the end of its life as a technical product. The debate ranges over whether people in the future will prefer to be able to 'down-load' any book they choose (via the Internet) and read it from an (ever more readable) screen; or whether they will still wish to 'curl up in bed' with a couple of hundred pages of printed paper inserted between two pieces of cardboard. If the future of the paper book is questionable, then that of the paper map is even more dubious. Paper maps have never been particularly 'user-friendly' documents to handle, as anyone who has tried to unfold one on a windswept mountainside, or has read a road atlas whilst driving, will testify. The map of the future will almost certainly not be drawn on paper. However, the contemporary alternatives of computer screen or hand-held monitor are not the only possible display technologies, as the discussion of virtual reality above indicates. One thing which can be predicted is that the product which indicates your position on a mountainside in future is likely to be quite different from that which helps you find the way while driving. There will be many different types of map in the future. This, more than any theoretical argument about the evolv-

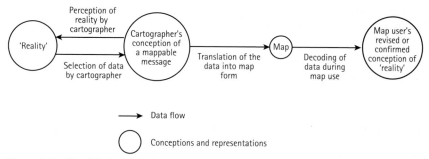

Figure 9.1   Simplified model of cartographic communication system.

ing nature of representation, will change our perspectives of what is 'cartograph-ically correct' for ever.

Changing perspectives on cartographic practice

Because maps are often now made by those who will use them, there is an increasingly 'private' dimension to the map user's interaction with the map arte-fact. The commonality of accepted practice in both map presentation (e.g. design parameters) and in map use (e.g. the 'correct' route in navigation always being the shortest, when it could be the cheapest, the most scenic or the most quiet) is being overwhelmed with a multitude of psychological, cultural and sociological views of what maps are and how they can be used as tools. This private interaction takes away from the map its previously perceived pre-eminent function as a communicator of information (from map-maker to map user), equivalent to the broadcast media as discussed on p. 157.

Much cartographic research in the 1970s and 1980s was directed towards the assessment of best practice in creating map artefacts that could most efficiently (i.e. with the least information loss) transmit the message or data which the car-tographer wished to proclaim to the recipient of the information, the map user. This view of cartography, as a communicative system (Figure 9.1) in which the cartographer extracts spatial data from the 'real world' (albeit under personal influence) and creates a well-defined message that is embodied in the graphical design of the map, is no longer regarded as appropriate in current cartographic theory. It assumes a 'knowledge engineering' approach, in which spatial data and their attributes can be packaged and transmitted with minimal alteration, and suggests that maps have a pre-defined purpose, which in most cases they patently do not. Current research focuses on a more flexible view of map-making and map use in a more self-contained domain, where the map-maker or data collector is also the map user. Thus, the map is now being created by a person interested in interacting with geographical data, with the prime purpose of facilitating investigation, exploring or confirming hypotheses, and improving decision-making, all by and for the individual (or small group) alone. The acceptance of the communication model by many cartographers led to a concentration of car-tographic research on the design elements of the map product (with the inten-

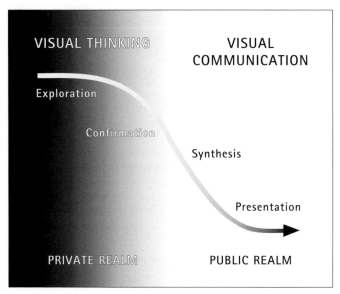

Figure 9.2   DiBiase's 1990 diagram of the realms of cartographic practice. Reproduced by kind permission of Elsevier Science Ltd.

tion of optimizing map design in order to optimize the communication of information). Today, 'post-modern' cartography gives more attention to other aspects of the mapping and map-making process – the decisions involved in data collection and compilation, the influence of the mental model of the map-maker and map user, and the importance of the task for which the map is being used in the processes of visualization and decision-making – and acknowledges differing assumptions, i.e. that the map is a subjective, not an objective, representation, and that artistry has a role to play in such representation. Such a view is not necessarily novel; indeed it could be seen emerging from the dominant post-war school of cartography. What is new now is its widespread acceptance.

'For maps of larger scale, an artistic objective might well lessen our insistence on a strict geometric framework for maps and make room for the greater use of mental constructs of social, cultural and economic space' (Robinson, 1989, p. 97). This does not mean that the map no longer has a communicative function, as in many cases the hypothesis or decision which the individual is investigating needs to be confirmed or displayed to a wider audience. Figure 9.2 indicates the relationship between these stages of map handling, depicting the 'private' and 'public' realms of cartography. The former involves a high level of interaction with the map: exploring the data, suggesting and confirming hypotheses and recording conclusions. The latter aspect is the (up to now) familiar picture of the map user reacting to a map produced by a distant, probably specialist cartographer, in response to a particular problem needing to be solved or task needing to be undertaken. Clearly, the communicative role of maps is primarily in the public realm, whilst the term 'cartographic visualization' has been given to the use of maps and spatial data in the more private realm. The distinction

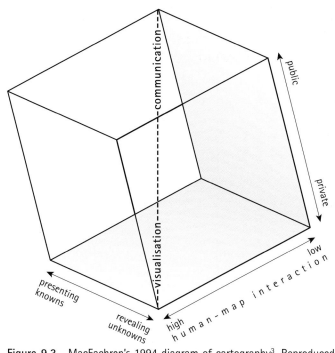

Figure 9.3   MacEachren's 1994 diagram of cartography[3]. Reproduced by kind permission of Elsevier Science Ltd.

between these two is not, of course, distinct and the axes of the cube constructed in Figure 9.3 develop the relationship between them.

Cartographic visualization

In order for maps to work efficiently within the domains outlined in Figure 9.3 (e.g. to undertake the range of map use tasks illustrated in Table 9.1) we must move beyond a perception of the map as a static two-dimensional object which portrays a message created by an active map-maker and passively perceived by a map user. New technology and new ways of thinking about society allow us to consider the representation of world-views in radically different ways from such a perception. The map user is increasingly likely to control the representation: interactive changes in content and the method of portrayal of information are useful and nowadays commonplace modifications to the previously linear model of cartographic production, embodied in the communication model explained above, which excluded any alteration of the appearance of the map once it had been compiled. Not only are we likely to want to change the map image to suit our requirements, but we may also want to actively interrogate the map face, ascertaining attributes at locations, determining position, and questioning real or implied relationships which are evident on the map. It may well be necessary to quickly create new and different ways of representing the data, not as maps but in the form of graphs, tables, animated sequences and supporting video

Table 9.1 MacEachren's 1994 exposition of map use tasks based on Figure 9.3 (after MacEachren and Taylor, 1994)

	High interaction		Low interaction	
	Revealing unknowns	Presenting knowns	Revealing unknowns	Presenting knowns
Private	Use of exploratory data analysis to compare alternative methods of presentation (e.g. of differing class intervals on choropleth maps)	Use of hypermedia tools to access a map collection	Use of 'graphic scripts', created in response to a 'user profile' to automatically navigate through a geographic data set	Use of a reference or inventory map to retrieve information (e.g. about the location, size and easements of a land parcel)
Public	Use of the Internet to give groups of researchers shared access to interactive simulations	Use by a TV meteorologist of a weather map 'chalkboard' onto which annotations (e.g. weather fronts) can be drawn	Use of a publicly available digital terrain model to explore the nature of the landscape	Use of a 'you are here' type location map for tourists in a city centre

clips. Pictures of the people living in certain places, and sound excerpts of some of their views of the world can now be integrated around what was the traditional map. The linkages between map and other computer-held data can be explicitly shown and easily used.

Contemporary computer-assisted mapping systems capable of fulfilling such requirements are called 'cartographic visualization systems' and they build on the concepts introduced in Chapter 6. The existence of such tools has led to a fundamental reassessment of map use, extending it into areas which previously did not routinely use maps, and has also, therefore, led to a reappraisal of the way in which maps work. It has obviously affected cartographic practice, allowing non-professional cartographers to become as skilled as trained map-makers in the techniques and methods of map production. This in turn has extended map use to communities which had previously had the maps of their world delivered to (or imposed on) them. The line between the professional and the non-professional worker in cartography has become blurred, as it has in many other walks of life. Rather than being quite incompetent in certain tasks, as was the case previously, people are increasingly more likely, willing and able to perform such tasks moderately well and be satisfied with their efforts. It is ironic that, for all its own perceived and well-documented mystique, information technology is removing much of the mystique that map-making used to hold. What drives this process forward most quickly is that it is children who find the new technology most easy to adopt, as to them it is merely one factor in their lives, among so many others, which is new yet to which they are receptive.

Map-makers of the future

Despite the fact that the International Cartographic Association (ICA) represents the profile of 'corporate cartography' at its most institutionalized and internationalized (i.e. highly organized, restricted, ordered as a service industry and serving the 'public'), even this body recognizes the importance of the individualistic, private and, spontaneous mapping urge in the cartographers of tomorrow.

The ICA's Children's Map Competition (The Barbara Petchenik Award: see Box 9.1) was initiated in 1993 and is open to all children under the age of 16. The 1995 competition attracted 120 entrants who constructed their view of the world with such varying titles as 'Wonderful world of a child's dream', 'The world is for everyone' and 'Children unite'. The images created by the winning children are used by the United Nations Children's Fund (UNICEF) in educational and promotional ways, having been incorporated into UNICEF wall posters and a United Nations CD-ROM called 'My City', and presented on the ICA home pages on the World Wide Web.

Even when they mature, it is unlikely that tomorrow's American cartographers will produce maps like those of Arthur Robinson or Barbara Petchenik. The view of the world held by children in the 1990s is very different from that held in the 1930s or 1960s. Not only is there a feeling that there is now more freedom over what can be mapped (partly due to the falling costs of mapping

Box 9.1   Personality box – Barbara Bartz Petchenik

Barbara Bartz Petchenik (1939–1992) had a wide-ranging career in both academic and commercial cartography. Having started her university studies (at the University of Wisconsin, USA) as a chemist, she was sufficiently attracted by the high profile of cartography in that institution to undertake postgraduate study in the field. The doyen of American cartographers, Arthur Robinson, whose work has influenced governmental, commercial and academic cartography throughout the USA since the 1940s, supervised her research work on typography, cartography and the use of text on maps, and collaborated with her in the writing of a seminal set of essays, published as *The Nature of Maps* in 1976. This profound book gives an insight into the prevalent cartographic research paradigm of the period.

Her major research interests were in map design, education and cognitive psychology. Such concerns led her naturally to the investigation of children's map use practices and in commemoration of her concern for this subject, and her period as a vice-president of the International Cartographic Association (ICA), the children's map competition, organized by the ICA in conjunction with UNICEF, has been entitled The Barbara Petchenik Children's Map Competition.

But Petchenik was also a practical map-maker. From 1964 to 1970 she was the cartographic editor and research and design consultant for the *World Book Encyclopaedia* and she continued in a similar vein for a further five years with the *Atlas of Early American History*. With her high profile within academic cartography, the publishing industry, the American Cartographic Association and the ICA, Barbara Petchenik blazed a trail which many women cartographers are destined to follow.

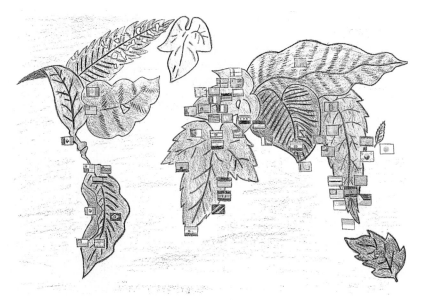

A winner of the ICA Petchenik Children's Map Competition 1993.

described above), but there has also been a change in the emphasis on what it is important to map. Thus, maps now being produced of California may highlight the sparse and diminishing sources of water in that state as environmental concerns grow, or the distribution of illegal immigrants in line with the rise in xenophobia. In the recent past the freeway system would dominate such maps partly because of its functional use but also as a symbol of American achievement. Even earlier, the maps would highlight the locations of resources – land to be grazed, hills to be mined. Just as California has changed from a state containing land ready to be exploited, to later become the richest and most developed part of the globe and now to a site for environmental degradation and political turmoil, so too are the maps of that part of the world altering. This process is still largely uneven however: the new maps of California are considerably more likely to be drawn by students in Berkeley than by young adults of the same age living in down-town Los Angeles. Despite this, the current changes perhaps reflect more the mapping of different times than the mapping of different peoples.

Mental maps

The importance of different groups' varying perceptions of the same spaces was brought to the attention of geography most clearly through the study of mental maps, which was an important subject of interest in the 1960s and 1970s. Figure 9.4 shows three views of Los Angeles: one from the perspective of white residents living in Westwood; the second from black residents of Avalon and the third through the eyes of Hispanic residents in Boyle Heights. They are three very different representations of one city. What is most important for our argument is that since the initial publication of work such as this, by university researchers, this particular concept has spread so that it is now common for students and schoolchildren to be taught that there are many different views of the world and that each can be equally valid. The change in teaching practices in general that has taken place in much of the western world since the 1960s has meant that today's children are less likely to feel that there is a correct way to draw a map which they must follow. Use of their imagination is encouraged more now than it was in the teaching of geography in the past (and in teaching, and growing up in general), so that along with the increased practical freedom which new technology has brought to map production, there has been an increasing freedom in what is permissible as well as feasible to draw.

Although the ICA Children's Map Competition has sought to draw examples of children's maps from around the world, it is striking to see, at each biennial conference of the ICA (where the entries are displayed), the world-wide distribution of children who have entered. The USA and UK provide the largest number of entrants, while from many countries only one example is submitted and from most countries in the world no maps are forthcoming. Although all children have views of the world to relay, their chance of being able to show their representations depends largely on where in the world they are born. Children (and the elderly) are also dominant amongst the numbers of the world's poor

(a)

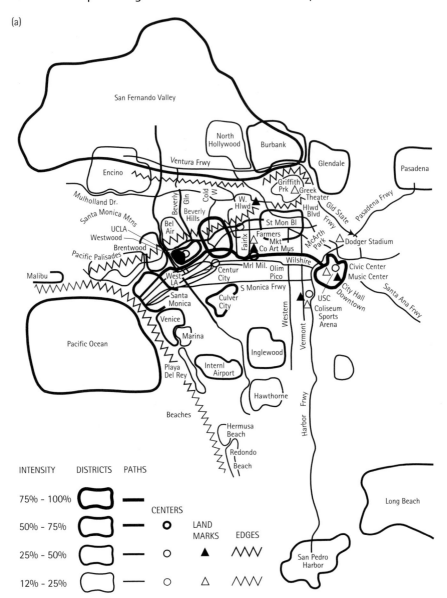

Figure 9.4 Mental maps of Los Angeles: (a) white residents living in Westwood.

and so the proliferation of new technology is unlikely to reach them, or their children or grandchildren, even at current rates of diffusion. It is important to remember that in one sense the democratization of mapping has meant that instead of only one person in every 100 000 drawing our maps of the world, now one person in 1000 is able to create and distribute their representation: a 100-fold rise in a very short time, but still only a very small proportion of the planet's population has been affected.

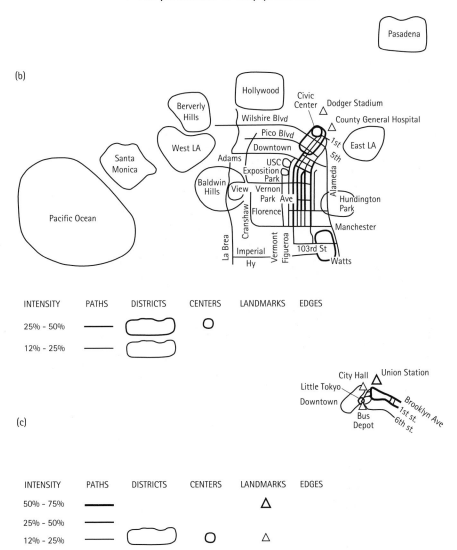

Figure 9.4 Mental maps of Los Angeles: (b) black residents living in Avalon; (c) Hispanic residents living in Boyle Heights.

The parameters of map production

In the light of recent changes in our views of people's perceptions and mappings of the world, in the technology used to transform these into map form, in the nature of the products which can carry such representations, in the societal and institutional framework within which map-making is undertaken and in the uses to which maps are put, it would seem foolhardy to make a definitive statement of future trends in this field. What we can say is that such changes show no sign of abating and the fluidity of this area of human endeavour will remain high. The nature of mapping and of what is mapped is likely to alter dramatically as a very different set of people is now drawing maps with very different tools. However, this development is likely to be undertaken within a context of uneven access to the means of production of maps and may even, for a short time, exacerbate the western-centric view of the world which most world mapping conveys. One way of looking at the spread of the Internet, and the growing numbers of graphical and map products it carries, is to see it as similar to the world-wide spread of the automobile, blue jeans and Coca-Cola – a diffusion of dominating American culture. Although more and more people are now signed on to the World Wide Web, what most of them are currently looking at are images created in North America by white, male, middle-aged and middle-class North Americans. This may well soon change, but until it does some degree of scepticism of the new technology and the products it disseminates is understandable.

There are other constancies in the changes which can now be seen, reflected in the parameters which impact on map creation, production and use and which we can profitably examine to gain a picture of the possible future of cartography. The notion of a 'world-view', which has been emphasized throughout this book, implies a recognition of the international nature of mapping and map-making with some parts being more international than others. Haggett's classic representation (Figure 9.5) of the world of a man's grandfather, his father and himself reveals the increasingly cosmopolitan nature of our world-view. At the same time the rise in international migration towards the West has meant that more and more 'alternative' world-views are being presented in the culturally dominant parts of the world, and these often conflict with preconceived perceptions extant in these areas. On the technical side, opportunities and requirements to travel for work, business and leisure, along with an increasing recognition of the spatial nature of human management and decision-making practices, have led to a burgeoning demand for spatial data products. Allied with the growth in technical skills evident in the general population and alluded to above, map-making is now being undertaken more often and by more people than ever before. The contemporary technology used to create map products allows for speedier production. Interest in the nature of mapping and map-making is also increasing, as evidenced by the recognition of the importance of 'graphicacy' (an appreciation of the nature of graphic design, in particular its relation to map production and use) and of the work of writers such as Tufte, who argue for a more competent awareness by all of the fundamentals of graphic design (see Further reading).

These changes all impact on map creation, production and use. As our world changes, our representations of it change too, and they in turn will change the way we all perceive how our places on this planet are organized and how we choose to organize them. The analytical, explanatory and monitoring roles of geography mentioned in the Introduction to this book will continue to require mapping and map-making, but the dividing line between maps-proper and map-like objects is likely to blur even further. One day, perhaps not too distant, today's paper maps will be studied with the same curiosity that the South Sea Islanders' ancient stick and shell constructions currently inspire. How did people ever find their way using those?

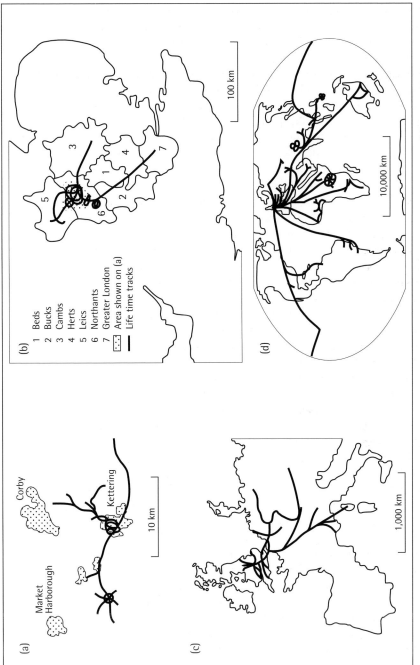

Figure 9.5 The world-views of different generations in the twentieth century: lifetime migration patterns of (a) the subject's great grand-father; (b) his grandfather; (c) his father; and (d) himself.

Further reading

The Open University Shape of the World series (1995) includes J. Allen and D. Massey, *Geographical Worlds*; J. Allen and C. Hamnett, *A Shrinking World*; and P. Sarre and J. Blunden, *An Overcrowded World?* The quotation on indigenous mapping was taken from D. Massey, 'Imagining the World', Chapter 1 in Book 1. E. R. Tufte has written two widely admired books on information display in general, which, in passing, consider the unique role of the map in particular: *The Visual Display of Quantitative Information* (1983) and *Envisioning Information* (1990), both published by Graphics Press, Chesire, Connecticut. The many issues relating to cartographic visualization are covered in A. M. MacEachren and D. R. F. Taylor, *Visualization in Modern Cartography* (Pergamon, Oxford, 1994). A more theoretical investigation of cartography, which covers the development of its research paradigms, and considers the role of visualization and its psychological and map use basis, is the wide-ranging *How Maps Work*, by A. M. MacEachren (Guilford Press, New York, 1995). Still of value and offering many valid insights into the relationships among cartography, psychology and language is A. H. Robinson and B. B. Petchenik, *The Nature of Maps* (University of Chicago Press, Chicago, 1976). The chapter by Robinson, 'Cartography as an Art', in D. W. Rhind and D. R. F. Taylor's *Cartography Past, Present and Future* (Elsevier Applied Science for the ICA, 1989, pp. 91–102) gives an interesting perspective on the artistic factors in cartographic design. From a practical viewpoint, guidance on how to use cartography to prepare maps of various types for a range of different purposes is given in M. Monmonier, *Mapping it Out: Expository Cartography for the Humanities and Social Sciences* (University of Chicago Press, Chicago, 1993).

References

This references section includes all material referenced throughout this book and further useful references.

Aberley, D., 1993, *Boundaries of Home: Mapping for Local Empowerment*, Gabriola, British Columbia: New Society Publishers.

Ackroyd, N. and Lorimer, R., 1994, *Global Navigation: A GPS User's Guide*, 2nd edition, London: Lloyd's of London Press.

Akerman, J., 1993, Blazing a well worn path: cartographic commercialism, highway promotion and automobile tourism in the United States, 1880–1930, *Cartographica*, **30** (1).

Allen, J. and Hamnett, C., 1995, *A Shrinking World*, Oxford: Oxford University Press.

Allen, J. and Massey, D., 1995, *Geographical Worlds*, Oxford: Oxford University Press.

Allen, J. P. and Turner, E. J., 1988, *We the People: An Atlas of American Ethnic Diversity*, New York: Macmillan Publishing Company.

Anonymous, 1989, The case against rectangular world maps, *The Cartographic Journal*, **26** (2), 156–157.

Bagrow, L. (revised and edited in translation by R. A. Skelton), 1964, *The History of Cartography*, Chicago: Precedent.

Barber, P. and Board, C., 1993, *Tales From the Map Room*, London: BBC Books.

Batty, M., 1995, The computable city, in Wyatt, R. and Gossain, H. (eds), *Proceedings of the 4th International Conference on Computers in Urban Planning and Urban Management*, Melbourne, Australia.

Batty, M. and Longley, P., 1994, *Fractal Cities*, London: Academic Press.

Beniger, J. R. and Robyn, D. L., 1978, Quantitative graphics in statistics: a brief history, *The American Statistician*, **32**, 1–11.

Blakemore, M. and Harley, J. B., 1980, *Concepts in the History of Cartography*, Cartographica Monograph 26, Toronto: University of Toronto Press.

Bricker, C., 1969, *A History of Cartography*, London: Thames and Hudson.

Brown, L. A., 1977, *The Story of Maps*, New York: Dover.

Bunge, W., 1988, *The Nuclear War Atlas*, Oxford: Blackwell.

Buttenfield, B. and Mackaness, W. 1991, Visualization, in Maguire, D. J., Goodchild, M. F. and Rhind, D. W. (eds), *Geographical Information Systems*, Vol. 1, Harlow, Essex: Longman.

Buttenfield, B. and McMaster, R., 1991, *Map Generalization: Making Rules for Knowledge Representation*, Harlow: Longman.

Canters, F. and Decleir, H., 1989, *The World in Perspective*, Chichester: John Wiley.

Chaliand, G. and Rageau, J. P., 1986, *The Strategic Atlas of World Geopolitics*, London: Penguin.

Cloke, P., Philo, C. and Sadler, D., 1991, *Approaching Human Geography: An Introduction to Contemporary Debates*, London: Chapman.

Cole, D., 1993, One cartographic view of American Indian Land Areas, *Cartographica*, **30** (1).

Coppock, T. and Rhind, D. W., 1991, The history of GIS, in Maguire, D. J., Goodchild, M. F. and Rhind, D. W. (eds), *Geographical Information Systems*, Vol. 1, Harlow, Essex: Longman.

Cosgrove, D. and Daniels, S. (eds), 1988, *The Iconography of Landscape*, Cambridge: Cambridge University Press.

Dangermond, J., 1992, What is a geographical information system?, in Johnson, A. I., Pettersson, C. B. and Fulton, J. L. (eds), *Geographic Information Systems and Mapping – Practices and Standards*, Philadelphia: American Society for Testing and Materials.

Dent, B. D., 1993, *Cartography: Thematic Map Design*, 3rd edition, Dubuque, Iowa: W. C. Brown.

Dorling, D., 1995, *A New Social Atlas of Britain*, Chichester: Wiley.

Drury, S., 1990, *A Guide to Remote Sensing: Interpreting Images of the Earth*, Oxford: Oxford University Press.

Edney, M., 1992, J. B. Harley (1932–1991): Questioning maps, questioning cartography, questioning cartographers, *Cartography and Geographic Information Systems*, **19** (3), 175–178.

Edney, M., 1993, The patronage of science and the creation of imperial space: the British mapping of India, 1799–1843, *Cartographica*, **30** (1).

Fisher, P (ed.), 1995, *Innovations in GIS 2*, London: Taylor and Francis.

Forster, S., Mommsen, W. J. and Robinson, R. (eds), 1988, *Bismarck, Europe and Africa*, Oxford: Oxford University Press.

Fothergill, S. and Vincent, J., 1985, *The State of the Nation: An Atlas of Britain in the Eighties*, London: Pluto Press.

Gilbert, M., 1982, *The Atlas of the Holocaust*, London: Michael Joseph.

Gleick, J., 1993, *Chaos: Making a New Science*, 2nd edition, London: Abacus.

Goodchild, M. F., 1988, Stepping over the line: technological constraints and the new cartography, *The American Cartographer*, **15** (3), 311–319.

Goodchild, M. F. and Gopal, S. (eds), 1989, *The Accuracy of Spatial Databases*, London: Taylor and Francis.

Goss, J., 1993, *The Map Maker's Art*, London: Studio Press.

Gould, P. and White, R., 1994, *Mental Maps*, Massachusetts: Allen & Unwin.

Greeves, T., 1987, *Parish Maps: Celebrating and Looking After Your Place*, London: Common Ground.

Hall, S., 1992, *Mapping the Next Millennium*, New York: Random House.

Harley, J. B., 1975, *Ordnance Survey Maps: A Descriptive Manual*, Southampton: Ordnance Survey.

Harley, J. B., 1988, Maps, knowledge and power, in Cosgrove, D. and Daniels, S. (eds), *The Iconography of Landscape*, Cambridge: Cambridge University Press.

Harley, J. B., 1989, Deconstructing the map, *Cartographica*, **26** (2), 1–20.

Harley, J. B., 1990, Cartography, ethics and social theory, *Cartographica*, **27** (2), 1–23.

Harley, J. B. and Woodward D., 1987, *The History of Cartography, Volume 1, Cartography in Prehistoric, Ancient and Medieval Europe and the Mediterranean*, Chicago: University of Chicago Press.

Harley J. B. and Woodward D., 1992, *The History of Cartography, Volume 2, Book 1, Cartography in the Traditional Islamic and South Asian Societies*, Chicago: University of Chicago Press.

Harley, J. B. and Woodward D., 1994, *The History of Cartography, Volume 2, Book 2, Cartography in Traditional East and Southeast Asian Societies*, Chicago: University of Chicago Press.

Hearnshaw, H. and Unwin D. (eds), 1994, *Visualisation in Geographical Information Systems*, Chichester: Wiley.

Henwood, D., 1994, *The State of the USA Atlas: The Changing Face of American Life in Maps and Graphics*, New York: Penguin.

Her Majesty's Treasury, 1994, *An Introduction to Geographic Information Systems*, London: Central Computer and Telecommunications Agency/HMSO.

Hodgkiss, A. G., 1973, The Bildkarten of Herman Bollman, *Canadian Cartographer*, **10** (2), 133–145.

Holmes, N., 1992, *Pictorial Maps*, London: Herbert Press.

Huntington, 1924, cited in Cloke *et al.* (1991).

ICA Gender and Cartography Working Group (1995), Directory of Women in Cartography, Surveying and GIS, International Cartographical Association.

Imhof, E., 1982, *Cartographic Relief Presentation*, Berlin: Walter de Gruyter (published in German in 1965).

International Cartographic Association, 1995, *Achievements of the ICA, 1991–95*, Paris: Institut Géographique National.

Johnson, A. I., Pettersson, C. B. and Fulton J. L. (eds), *Geographic Information Systems and Mapping – Practices and Standards*, Philadelphia: American Society for Testing and Materials.

Kain, R. and Baigert, E., 1992, *The Cadastral Map in the Service of the State: A History of Property Mapping*, Chicago: University of Chicago Press.

Keates, J. S., 1989, *Cartographic Design and Production*, 2nd edition, Harlow: Longman.

Keates, J. S., 1996, *Understanding Maps*, 2nd edition, Harlow: Longman.

Kidron, M., 1995, *The State of the World Atlas*, 5th edition, London: Penguin.

Kidron, M. and Segal, R., 1981, *The State of the World Atlas*, London: Pan Books (Heinemann).

Kidron, M. and Segal, R., 1984, *The New State of the World Atlas*, London: Pluto Press.

Kidron, M. and Smith, D., 1983, *The War Atlas: Armed Conflict – Armed Peace*, London: Pluto Press.

Kidron, M. and Smith, D., 1991, *New State of War and Peace: An International Atlas*, London: Grafton Books.

Kidron, M. and Smith, D., 1995, *The State of the World Atlas*, 5th edn, London: Penguin.

Konvitz, J., 1987, *Cartography in France 1660–1848*, Chicago: University of Chicago Press.

Kraak, M. and Ormeling, F., 1996, *Cartography: Visualisation of Spatial Data*, Harlow: Longman.

Krygier, J., 1996, Geography and cartographic design, in Wood, C. H., and Keller, C. P. (eds), *Cartographic Design: Theoretical and Practical Perspectives*, Chichester: Wiley, Ch. 3.

Larsson, G., 1991, *Land Registration and Cadastral Systems*, Harlow: Longman.

Lewis, G. M., 1987, The origins of cartography, in Harley, J. B. and Woodward, D. (eds), *The History of Cartography, Volume 1, Cartography in Prehistoric, Ancient and Medieval Europe and the Mediterranean*, Chicago: University of Chicago Press.

Livingstone, D., 1993, *The Geographical Tradition: Episodes in the History of a Contested Enterprise*, Oxford: Blackwell.

MacEachren, A. M., 1994, *Some Truth with Maps*, Washington: Association of American Geographers.

MacEachren, A. M., 1995, *How Maps Work*, New York: Guildford Press.

MacEachren, A. M. and Taylor, D. R. F. (eds), 1994, *Visualization in Modern Cartography*, Oxford: Pergamon.

Maguire, D. J., Goodchild, M. F. and Rhind, D. W. (eds), 1991, *Geographical Information Systems. Volume 1: Principles. Volume 2: Applications*, Harlow: Longman.

Maling, D. H., 1989, *Measurement from Maps*, Oxford: Pergamon.

Maling, D. H., 1993, *Coordinate Systems and Map Projections*, 2nd edition, Oxford: Pergamon.

Mandelbrot, B., 1983, *The Fractal Geometry of Nature*, San Franciso: Freeman.

Massey, D., 1995, Imagining the world, in Allen, J. and Massey, D. (eds), *Geographical Worlds*, Oxford: Oxford University Press.

McCormick, B. H., DeFanti, T. A. and Brown, M. D. (eds), 1987a, Visualisation in Scientific Computing, Special Issue of *Computer Graphics*, **21** (6), New York: ACM SIGGRAPH.

McCormick, B. H., DeFanti, T. A. and Brown, M. D. (eds), 1987b, Visualisation in Scientific Computing, Synopsis, *IEEE Computer Graphics and Applications*, 7, 61–70.

McLaren, R. A. and Kennie J. J. M., 1989, Visualization of digital terrain models: techniques and applications, in Roper, J. F. (ed.), *Three Dimensional Applications in Geographical Information Systems*, London: Taylor and Francis.

McNab, A., 1993, *Bravo Two Zero*, London: Corgi.

Mesev, T. V., Longley, P. A., Batty, M. and Xie Y., 1995, Morphology from imagery: detecting and measuring the density of urban land use, *Environment and Planning A*, 27, 759–780.

Monmonier, M., 1991, *How to Lie with Maps*, Chicago: University of Chicago Press.

Monmonier, M., 1993, *Mapping it Out: Expository Cartography for the Humanities and Social Sciences*, Chicago: University of Chicago Press.

Monmonier, M., 1995, *Drawing the Line*, New York: Henry Holt and Company.

Mounsey, H. (ed.), 1988, *Building Databases for Global Science*, London: Taylor and Francis.

Nicholson, N. L. and Sebert L. M., 1981, *The Maps of Canada*, Folkestone: Dawson.

Openshaw, S. (ed.), 1995, *The Census Users' Handbook*, Cambridge: GeoInformation.

Orlove, B., 1993, The ethnology of maps: the cultural and social contexts of cartographic representation in Peru, *Cartographica*, **30** (1).

Ormeling, F. J. Snr, 1987, In Memory, Eduard Imhof, *The Cartographic Journal*, **24** (1), 83.

Owen, T. and Pilbeam, E., 1992, *Ordnance Survey: Map Makers to Britain Since 1791*, London: HMSO.

Pickles, J. (ed.), 1995, *Ground Truth: The Social Implications of Geographic Information Systems*, New York: Guilford.

Pred, A., 1984, Places as a historically contingent process: structuration and the time-geography of becoming places, *Annals of the Association of American Geographers*, 74 (2), 279–297.

Pred, A., 1986, *Place, Practise and Structure: Social and Spatial Transformation in Southern Sweden: 1750–1850*, Totowa, New Jersey: Barnes and Noble.

Rabenhorst, T. and McDermott, P., 1989, *Applied Cartography: Introduction to Remote Sensing*, Columbus, Ohio: Merrill.

Raper, J. F. (ed.), 1989, *Three Dimensional Applications in Geographical Information Systems*, London: Taylor and Francis.

Ravenhill, W., 1992, Obituary, John Brian Harley, 1932–1991, *Transactions of the Institute of British Geographers*, New Series, 17, 363–369.

Rhind, D. W., 1988, Personality as a factor in the development of a new discipline: the case of computer assisted cartography, *The American Cartographer*, **15** (3), 277–289.

Rhind, D. W. and Taylor, D. R. F., 1989, *Cartography Past, Present and Future*, London: Elsevier Applied Science for the ICA.

Ritchie, W., Tait, D., Wood, M. and Wright, R., 1988, *Surveying and Mapping for Field Scientists*, Harlow: Longman.

Robinson, A., 1989, Cartography as an art, in Rhind, D. W. and Taylor, D. R. F. (eds), *Cartography Past, Present and Future*, London: Elsevier Applied Science/ICA.

Robinson, A. H. and Petchenik, B. B., 1976, *The Nature of Maps*, Chicago: University of Chicago Press.

Robinson, A., Morrison, J., Muehrcke, P., Kimerling, A. and Guptill, S., 1995, *Elements of Cartography*, 6th edition, Chichester: John Wiley.

Rundstrom, R. (ed.), 1993, Introducing cultural and social cartography, *Cartographica* **30** (1), Toronto: University of Toronto Press.

Sarre, P. and Blunden, J., 1995, *An Overcrowded World?*, Oxford: Oxford University Press.

Seager, J. and Olson, A., 1986, *Women in the World: An International Atlas*, London: Pan Books.

Seeber, G., 1993, *Satellite Geodesy: Foundations, Methods and Applications*, Berlin: W. de Gruyter.

Shawcross, W., 1986, *Sideshow: Kissinger, Nixon and the Destruction of Cambodia*, London: Hogarth Press, London

Snyder, J., 1993, *Flattening the Earth*, Chicago: University of Chicago Press.

Szegö, J., 1984, *A Census Atlas of Sweden*, Stockholm: Swedish Council for Building Research.

Szegö, J., 1987, *Human Cartography*, Stockholm: Swedish Council for Building Research.

Szegö, J., 1994, *Mapping Hidden Dimensions of the Urban Scene*, Stockholm: Swedish Council for Building Research.

Taylor, P. J., 1990, Editorial comment: GIS, *Political Geography Quarterly*, **9** (3), 211–212.

Tufte, E. R., 1983, *The Visual Display of Quantitative Information*, Chesire, Connecticut: Graphics Press.

Tufte, E. R., 1990, *Envisioning Information*, Chesire, Connecticut: Graphics Press.

Unwin, D. and Hearnshaw, H. (eds), 1994, *Visualisation in Geographical Information Systems*, Chichester: Wiley.

Williams, J. E. D., 1994, *From Sails to Satellites*, Oxford: Oxford University Press.

Wolf, P. and Brinker, R. C., 1993, *Elementary Surveying*, 9th edition, Hinsdale, Illinois: HarperCollins.

Wood, C. H. and Keller, C. P. (eds), 1996, *Cartographic Design: Theoretical and Practical Perspectives*, Chichester: Wiley.

Wood, D., 1992, *The Power of Maps*, New York: Guilford Press.

Wood, D., 1993, What makes a map a map, *Cartographica*, **30** (1), 81–86.

Wood, M., 1984, The panoramic map of Central Scotland, *Bulletin of the Society of University Cartographers*, **17** (1), 1–7.

Index